Comparative Pathobiology

Volume 3
INVERTEBRATE
IMMUNE RESPONSES

Comparative
Pathobiology

Comparative Pathobiology

Volume 3
INVERTEBRATE
IMMUNE RESPONSES

Edited by Lee A. Bulla, Jr.

United States Department of Agriculture
Manhattan, Kansas

and

Thomas C. Cheng

Lehigh University
Bethlehem, Pennsylvania

Springer Science+Business Media, LLC

Library of Congress Cataloging in Publication Data

Main entry under title:

Invertebrate immune responses.

(Comparative pathobiology; v. 3)
"Proceedings of a symposium held at Oregon State University, Corvallis,
on August 16-22, 1975. . . co-sponsored by the Society for Invertebrate
Pathology (SIP) and the American Society of Zoologists (ASZ)."
Includes index.
1. Invertebrates—Physiology—Congresses. 2. Immune response—Congresses. I. Bulla, Lee A. II. Cheng, Thomas Clement. III. Society for Invertebrate Pathology. IV. American Society of Zoologists. V. Series.
QL364.I58   592'.02'9   77-12385

ISBN 978-1-4615-7301-2   ISBN 978-1-4615-7299-2 (eBook)
DOI 10.1007/978-1-4615-7299-2

Acknowledgments

We thank Aileen Berroth, whose skillful attention to the typing and assembly of the material in this volume helped make the publication possible.

FOREWORD

This represents the third volume of the series entitled
Comparative Pathobiology. The chapters included represent the
proceedings of a symposium held at Oregon State University,
Corvallis, on August 16-22, 1975. The symposium was co-sponsored
by the Society for Invertebrate Pathology (SIP) and the American
Society of Zoologists (ASZ).

In recent years there has been an impressive increase in
interest in comparative immunology, i.e., a comparative approach
to understanding how animals, both vertebrates and invertebrates,
defend themselves against nonself materials. Ever since
Metchnikoff's pioneering studies during the late 1800s on the role
of phagocytes of invertebrates, which led to his theory of cellular
immunity, invertebrates have been employed with increasing fre-
quency for studying cellular defense. Consequently, it is not
surprising that included in the memberships of SIP and ASZ are a
large number of individuals with an active interest in this area
of research. As indicated by the chapters included in this
volume, the animal models employed have been primarily molluscs
and insects, although crustaceans and annelids have also been
popular.

Not only are invertebrates excellent models for studying cellular
immunity from the viewpoint of basic research, there are at least
two other reasons why the immune mechanisms of these animals are
being intensively examined: (1) With the increased employment of
aquaculture for the production of edible molluscs and crustaceans,
there is an urgent need to understand what infectious diseases
afflict these animals and how their bodies confront the causative
agents; and (2) in the development of biological control agents
against undesirable insects and molluscs, it is insufficient to
search empirically for microbial pathogens, we must understand how
the target organisms recognize self from nonself and react to the
latter. Even a potentially pathogenic microorganism would be ineffective as a control agent if it is recognized as nonself and
subsequently killed by the host's immune mechanisms.

By bringing together the papers comprising this volume, it is
our hope that another useful contribution has been made by
Comparative Pathobiology.

In conclusion, we would once again like individuals who have an interest in comparative pathobiology to contact either of us if they are planning to organize significant symposia. The pages of this series could serve as the publication medium.

Lee A. Bulla, Jr.

Thomas C. Cheng

Editors

Preface

   The collected papers on various aspects of invertebrate immunity
that were presented at this symposium are of high quality. Although
one dominant theme is phagocytosis, an old observation, the chapters
produced for this volume show imagination and breadth of experimental and technical approaches. Indeed, invertebrate immunology
is beginning to move in directions that go further than 19th
century scientists such as Haeckel and Metchinkcff. Phagocytosis
is a universal event characteristic of protozoans and populations
of macrophages or their precursors in metazoans. Thus, it alone
is probably the common event in all considerations of immunity.
This behooves present day investigators to do more than restate
that particular animals possess a group of cells that can phagocytose. Certainly, attempts to describe or discover the recognition unit on invertebrate phatocytes that enables them to ferret
out foreign material is of prime importance. Another meaningful
approach would be to determine if hemolymph components are derived from hemocytes and, if so, what role they play in immunity.
Since vertebrate immunity and its specificity is mediated by
antibodies, how then is humoral immunity, although apparently
lacking specificity by current methods, in invertebrates determined
and regulated?
   It is now known that earthworms, *Aporrectodea trapezoides,* can
encapsulate nematodes, as revealed in the paper by Poinar and
Hess. Nematodes are first surrounded by a uniform homogeneous
deposit that may be a humoral response, and that eventually
darkens suggesting a melanization process. Later, host amoebocytes
surround them forming what are known as brown bodies, host capsules
important in nonspecific defense reactions. This is unlike the
specific response of earthworms to tissue xenografts, however, the
last phases of graft rejection may include some features of brown
body formation that occur after grafts are recognized, reacted
against, and destroyed. In the case of nematode parasites or
foreign grafts, both are destroyed; the parasite by a process that
does not require the temporary acceptance of notself followed by
its destruction; the xenogeneic graft because it possesses
recognizable antigens that are first accepted as self, and then
destroyed.

By using biochemical and electron microscopic approaches, Cheng finds evidence for a double role of phagocytosis in molluscs, that of immunity and nutrition. We have long recognized that phagocytosis in protozoans is together an act of nutrition and defense, and here in a complex metazoan, Cheng clarified what is probably a common trait in the animal kingdom. He found that glycogen synthesis begins from bacterial constituents in molluscan phagocytes, first being detectable after 6 hr and reaching a peak at 24 hr. Glycogen is released from phagocytes after 16 hr post-injection and presumably it is utilized as an energy source once it is in the circulation. As a step further, Feng, Feng, and Yamasu found two types of amoebocytes (agranular and granular) in *Crassostrea gigas* and *Mytilus edulis*. These amoebocytes are capable of defense, obviously because of cytoplasmic organelles that manufacture, store and utilize enzymes, e.g., acid phosphatase. Amoebocytes play a role in nutrition as revealed by the use of certain xanthophylls as natural tracers for the transportation of food particles across the intestinal epithelium.

In their work on decapod crustaceans, Poinar and Hess found the formation of similar capsules to nematode parasites, however some are formed by muscle cells, the others by host blood cells. Decapod crustaceans have a cellular defense system similar to insects, in that their hemocytes react to metazoan parasites and, in addition, muscle cells may respond to the presence of parasitic nematodes. The hemocytic response is effective in halting the further development or it may even destroy parasites.

Schapiro and her group found that the American lobster, *Homarus americanus*, is ideal for studies of phagocytosis since it can efficiently clear almost all bacteria from its hemolymph. They used an *in vitro* system of hemocytes from *H. americanus* to assess the efficiency of phagocytosis of virulent and avirulent strains of *Aerococcus viridans* (formerly *Gaffkya homari*). Opsonization was necessary for phagocytosis to occur. The avirulent strain, when opsonized, was phagocytized by greater than 90% of the hemocytes. However, the virulent strain was phagocytized by almost 40% of the hemocytes using the same conditions and there were fewer virulent bacteria per phagocytic cell. Her experiments are clearly presented and their implications to the existence of receptors on invertebrate hemocytes are noteworthy. Finally, it is of obvious importance to problems of nutrition, since healthy lobsters are edible.

Vinson's work is of interest, for it offers a contrasting presentation to the usual condition - the foreign material's response to the host. He emphasizes mechanisms for the evasion of host's defenses by insect parasitoids. One view holds that depletion of the host's food reserves or its hemocytes may account for the protection of parasitoid larvae. The parasitoid egg may escape the host's defense because of the presence of a surface coating, possibly a mucopolysaccharide which is susceptible to degradation by the host. Obviously this problem is related to the mechanism of self recognition.

Nappi's paper is interesting because it deals with the hemocytes of *Drosophila* in two distinct situations – that of tumor formation and of encapsulation, a non-specific immune response. He acknowledges that although significant contributions have been made in studying insect immunity, there remain unanswered questions about the mechanisms of hemocyte activation and the specificity of their interactions with various infectious and oncogenic agents. Encapsulation of metazoan parasites may be similar to those producing abnormal cellular growths known as melanotic tumors in mutant strains of *Drosophila*. In tumor formation and encapsulation, hemocyte reactions involve a precarious mass differentiation of hemocytes which aggregate and adhere to each other to form pigmented capsules.

Anderson's work on phagocytosis is of much importance to those interested in comparative mechanisms of phagocytosis. As a result events which occur in invertebrate cells when compared with situation in humans, readily provides instances of applications of basic research to clinical problems. In mammals, neutrophils show marked oxygen utilization, hexose monophosphate pathway activation, H_2O_2 production, and the nitroblue tetrazolium reduction. None of these processes occurs in invertebrate phagocytes nor in most mammalian macrophages. Neutrophils kill bacteria via the myeloperoxidase $-H_2O_2-$halide system and lysosomal cationic proteins, mechanisms that are absent from invertebrate cells and mammalian macrophages. Anderson's other interesting findings are different. When confronted with formalized sheep erythrocytes, hemocytes adhere immediately to insect hemocytes and form rosettes; no phagocytosis occurs. Other particles such as yeast, bacteria and carmine also readily adhere and are seldom phagocytized. Glycolytic inhibitors (iodoacetate, arsenate, fluoride) substantially inhibit the phagocytosis of carbon. By using comparative metabolic studies of phagocytizing invertebrate hemocytes, we know something about the energy for phagocytosis in all animal species.

The essence of Chadwick's presentation is of much interest since she attempts to pull together volumnious amounts of information related to insect humoral immunity. She stresses however, that few investigators have exhaustively used new separation techniques to characterize the various humoral components induced in insects after immunization – an area that requires intensive investigation. Moreover she recognizes as others have in the past the need for understanding the relationship between humoral components and their source, presumably hemocytes. Finally she summarizes the general characteristics of induced immunity in insects, useful information. Insect immunity is rapidly acquired, a single stimulus is adequate, it is of brief duration, relatively non-specific, not associated with immunoglobulin and cell-free hemolymph shows antibacterial activity.

One cannot make a comparative study of invertebrate immune
responses without formulating general concepts on the nature of
the evolutionary processes. Most structural changes in cells of
the immune system are probably adaptive modifications to a variety
of environments and modes of life exhibited by invertebrates.
Clearly, there is not sufficient information from existing,
functional, comparative studies of diverse invertebrate immune
systems to even speculate on what the nature of the pressures
are that have brought about these evolutionary adaptations.
According to current systematics, the annelids, molluscs and
arthropods (protostomes) did not give rise to the vertebrate line.
Nevertheless, continued intense analyses of their immune systems
and those of echinoderms and tunicates that belong to the
deuterostomate group from which vertebrates arose, are necessary
if we are to devise meaningful concepts regarding the universality
of immune responsiveness. Studies of immunity as a biologic
phenomenon should not be restricted to searching for immune
responses that mimic those of vertebrates, particularly mammals.
Similarly, research on invertebrate immunity should not have as
its prime aim to understand how human immune reactions evolved.
A broader view holds that immunity is as ubiquitous as respiration,
reproduction, behavior, etc. and understanding its condition in
all animals broadens and unifies our knowledge.

E. L. Cooper

Department of Anatomy
University of California
  Medical Center
Los Angeles, California 90024

Contents

Biochemistry and Physiology of Invertebrate Macrophages In Vitro

ROBERT S. ANDERSON

SLOAN-KETTERING INSTITUTE FOR CANCER RESEARCH
DONALD S. WALKER LABORATORY
RYE, NEW YORK

I. INTRODUCTION

Phagocytosis is a cellular function which is commonly observed in all animals. In unicellular animals and in certain metazoans, phagocytosis serves as the major means of nutrient ingestion. At all phylogenetic levels, phagocytic cells seem to be capable of distinguishing between self and not self. In higher forms the predominant physiological role of phagocytosis has to do with cellular defense mechanisms. In many organisms both circulating and fixed phagocytes are active in clearing foreign particulates from extracellular fluids. A diverse array of particles can be phagocytized including viruses, bacteria, yeast, red blood cells, colloidal suspensions of many kinds, etc. Foreign bodies too large to be ingested by phagocytes are usually encapsulated by hemocytes and thus isolated from the host. The fate of ingested particles varies; however, one of two alternative mechanisms is frequently observed. In one case material subject to digestion by hydrolytic lysosomal enzymes is catabolized intracellularly. On the other hand, material which cannot be broken down within the phagocytes is carried to the periphery of many invertebrates and released into the external environment. This contrasts sharply with the fate of similar material in vertebrates in which considerable quantities of biologically inert material is stored for relatively long periods of time within the reticuloen-dothelial system.

This paper deals with phagocytosis by circulating invertebrate phagocytes. *In vitro* reactions are emphasized because in these studies the environment in which phagocytosis occurs can be controlled. Such *in vitro* systems permit the study of physio-logical aspects of hemocyte particle interactions such as the involvement of opsonins. We are also able to make a number of preliminary observations on the metabolism of phagocytizing hemocytes.

Both old and new information are integrated in this paper. This is not intended to be a detailed review; a more complete treat-ment of the literature is available (Anderson, 1975). Here only a few representative papers to illustrate each point are mentioned; also some very recent articles are discussed. Pre-sented for the first time are observations on the phagocytic potential of continuous hemocyte cell lines from *Estigmene acrea*. These cell lines were established by Dr. Robert R. Granados and co-workers, at the Boyce Thompson Institute, Yonkers, New York, who graciously provided a continuous supply of material for the phagocytic assays carried out in my laboratory.

II. OBSERVATIONS OF IN VITRO PHAGOCYTOSIS BY INVERTEBRATE CELLS

It is clear that acellular organisms phagocytize several classes of nonfood particulates. Githens and Karnovsky (1973) have studied the uptake of polystyrene (PS) beads by a slime mold *Polysphondylium pallidum*. During the first 15 min of incubation, almost all of the cells phagocytized at least one bead. Data on the kinetics of phagocytosis generated in this study were compared to those produced in similar studies of protozoans and mammalian polymorphonuclear cells (PMN). The slime mold cells had a greater affinity for the polystyrene beads than did PMN. Similar studies of PS uptake by *Acanthamoeba* had been carried out by Weisman and Korn (1967). The PS beads were bound to the cell surface and ingested after an optimal mass accumulated. The uptake kinetics were a function of total bead mass not size or number. Beautiful scanning electron micrographs of the details of PS bead ingestion by *Acanthamoeba castellanii* were made by Goodall and Thompson (1971). Rabinovitch and DeStefano (1971a) showed that erythrocytes (E) from various mammalian species were taken up to different extents by *Acanthamoeba*. Chemical treatment of the E with gluteraldehyde, formaldehyde, tannic acid, etc. caused augmented phagocytosis. Phagocytosis was temperature-sensitive, and attachment and ingestion could be disassociated at low temperatures (5-10°C). Many of the above characteristics of *in vitro* phagocytosis have been reported at all phylogenetic levels.

In vitro phagocytosis has been studied in annelids, mollusks, arthropods, echinoderms, and ascidians. As might be expected, the majority of studies concentrate on mollusks and arthropods. Bang (1961) described adhesion of bacteria to hemocytes of the American oyster, *Crassostrea virginica;* actual ingestion of bacteria was not always seen. The bacteria were frequently trapped in a fibrous network formed as a result of extracellular clotting. This mechanism was thought to serve to immobilize the bacteria prior to phagocytosis. Phagocytic hemocytes of many mollusks, including the snail, sea hare, and octopus, have been studied *in vitro*. Several of these studies will be mentioned in the following section on opsonins.

Phagocytosis of sheep E by crayfish *(Parachaeraps bicarinatus)* hemocyte monolayer cultures was studied by McKay and Jenkin (1970a). It was found that cells from animals "immunized" by bacterial vaccines were phagocytically hyperactive (McKay and Jenkin, 1970b). What relation this phenomenon has to macrophage activation, as it is known in mammals, has as yet to be established. Possibly,attempts to find phenomena in invertebrates comparable to vertebrate immunological reactions will be most fruitful in this area. Invertebrate "immune" responses are often nonspecific and short-lived; whereas mammalian immunity is characterized by specificity and anemnesis conferred by components of the

lymphoid system, which are not detectable in invertebrates. In
vertebrates, and possibly invertebrates, activated macrophages
show an increased, nonspecific phagocytic activity.

Insect cell cultures are capable of phagocytizing many kinds of
foreign particles. The phagocytic responses of an established
line of insect hemocytes will be discussed in another section
of this paper. Cytoplasmic polyhedrosis develops in cultured
Antheraea eucalypti cells after phagocytosis of polyhedra.
These cultures also took up starch grains and carbon suspensions
(Grace, 1962). Grace (1971) concludes that there is little
doubt that cultured insect cells can phagocytize viruses and
bacteria. The various stages of phagocytosis have been recorded
on film by Grace and Day (1963) and Vago (1964).

Rabinovitch and DeStefano (1970) carried out comparative
studies of phagocytosis by hemocytes of the wax moth, Galleria
mellonella, and mouse macrophages. Untreated and effete
erythrocytes were not extensively phagocytized in vitro.
However, treatment of E with formaldehyde, tannic acid, saponin,
colloidal silica, ferrous sulfate, and carbodiimide enhanced
their ingestion. In most cases these treated E were phagocytized
by both wax moth and mouse macrophages in the absence of hemolymph or serum, respectively. The authors suggested that common
features of the membranes of these phylogenetically distant
phagocytes might underlie recognition in this "nonimmunological"
phagocytosis.

Scott (1971a) studied phagocytosis by monolayer preparations
of hemocytes of the cockroach, Periplaneta americana. These
cells shared the mammalian macrophage characteristics of
thigmotaxis and neutral red ingestion. Sheep and chicken E
readily adhered to the cells; this was interpreted as the first
stage of phagocytosis. Actual ingestion of E was rarely seen,
possibly as a result of suboptimal experimental conditions.
The in vitro bactericidal capacity of hemocytes from another
cockroach, Blaberus craniifer, was studied by Anderson et al.
(1973a). In these short-term cultures it was shown that the
hemocytes were the sole mediators of bacterial killing. The
medium itself was not bactericidal, neither was hemocyte lysate
nor medium in which phagocytically stimulated or unstimulated
cells had been maintained. Cell-free hemolymph was not bactericidal
at the concentration present in the medium. If glycolytic
inhibitors, at concentrations not toxic to the bacteria and not
lethal to the cells, were present in the medium, no bactericidal
activity was recorded due to the inability of the cells to phagocytize (Anderson et al., 1973b). Microscopical examination of
cells from inhibitor-free cultures showed ingested bacteria,
often in well-developed phagocytic vacuoles. A recent study of
in vitro phagocytosis of bacteria by Galleria mellonella larval
hemocytes has shown that ingested Escherichia coli are not always
surrounded by a vacuole (Ratcliffe and Rowley, 1974). In this
paper the stages of bacterial ingestion are documented by

electronmicrography.

Several interesting groups such as the echinoderms and urochordates are found on the invertebrate evolutionary path leading to the vertebrates. Sea urchin hemocytes took up both Gram-positive and -negative bacteria *in vitro* (Johnson, 1969; Johnson *et al.*, 1970). This process was recorded in electron-micrographs. There are few reports of *in vitro* phagocytosis by tunicate blood cells. *Ascidia atra* hemocytes are reported to take up carmine *in vitro* (Fulton, 1920). Studies in this laboratory indicate that hemocytes from *Halocynthia pyriformis* can take up yeast particles and certain mammalian E.

III. EVIDENCE FOR OPSONIC INVOLVEMENT IN <u>IN VITRO</u> PHAGOCYTOSIS BY INVERTEBRATE HEMOCYTES

It is well known that mammalian macrophages more avidly phagocytize foreign substances opsonized with antibody and com-plement. The ability of invertebrate humoral factors to perform in the same way has been reported in a number of species. How-ever, in most cases invertebrate hemocytes can phagocytize unopsonized particles in the absence of hemolymph to at least a minimal extent. Exceptions will be noted below. Oponizing activity has been most frequently associated with invertebrate hemagglutinins (lectins). In contrast, the hemolymph of many species in which lectins are easily demonstrated has been shown not to be opsonic. Also, in other species lectin-free hemolymph will facilitate phagocytosis. In studies of the ingestion of several kinds of particles by hemocytes from a given species, it can be seen that hemolymph may influence the uptake of only certain particles. At this time it is unwise to make general statements concerning opsonization in invertebrates; reports must stand on their individual merit.

The presence of classical mammalian immunoglobulins has not been detected in any invertebrate. Macrophages from metazoan inverte-brates, not surprisingly, lack receptors for mammalian antibodies. Erythrocytes sensitized with rabbit antiserum were not phago-cytized to a greater extent than untreated E by cells from the lesser octopus, *Eledone cirrosa* (Stuart, 1968). Sensitized red cells were not ingested by hemocytes of the wax moth, *Galleria mellonella*, but were extensively phagocytized by mouse macro-phages (Rabinovitch and DeStefano, 1970). Sheep E sensitized with rabbit antiserum did not adhere to *Periplaneta americana* hemocytes but were avidly taken up by mouse peritoneal macro-phages. Curiously, mammalian antibodies promote phagocytosis by slime molds and protozoans. Smooth forms of *Salmonella* were phagocytized slowly, if at all, by *Dictyostelium discoideum*, unless they were opsonized with rabbit anti-*Salmonella* serum (Gerisch *et al.*, 1967). Rabinovitch and DeStefano (1971b) showed that antibody and plant lectins stimulated phagocytosis of sheep E by *Acanthamoeba*. Normal rabbit serum and immune serum adsorbed with sheep E were not opsonic.

Oyster, *Crassostrea virginica*, hemolymph contains a hemagglutinin directed against human ABO blood group antigens and several other vertebrate E (Tripp, 1966). Phagocytosis of rabbit E by oyster hemocytes proceeded in the absence of the hemagglutinin; however, phagocytosis was stimulated by the addition of hemagglutinin-containing shell liquor. This stimulation was reflected by increases in both the numbers of cells phagocytizing and the numbers of E/phagocyte. Further studies (Tripp and Kent, 1967) confirmed that the rate and extent of *in vitro* phagocytosis were stimulated by the presence of appropriate concentrations of lectin-containing hemolymph. Another mollusk, *Helix aspersa*, has been reported to possess an opsonin which is absolutely required for *in vitro* phagocytosis of formalized sheep E and formalized yeast (Prowse and Tait, 1969). Furthermore, the opsonin showed specificity in that no phagocytosis of either kind of particle occurred in serum absorbed with the homologous particle; however, phagocytosis occurred in serum absorbed with the heterologous particle and in normal serum. The exact nature of the opsonin has yet to be determined; however, many species of *Helix* contain naturally-occurring hemagglutinins.

A recent study of opsonins in another gastropod, *Otala lactea*, which lacks humoral lectins indicated that hemolymph factors could stimulate phagocytosis of formalized yeast (Anderson and Good, 1976). In this species, hemolymph did not increase phagocytosis of formalized sheep E or several kinds of bacteria. However, hemagglutinin (from extracts of *Otala* albumin gland) were shown to promote phagocytosis of formalized sheep E.

A serum factor agglutinating marine bacteria and vertebrate E was described by Pauley *et al.* (1971a) in the sea hare, *Aplysia californica*. Pauley *et al.* (1971b) also showed that serum stimulated *in vitro* phagocytosis of chicken E by *Aplysia* hemocytes. It was suggested that the lectin and the opsonin might well be the same. *Serratia marcescens* persisted in the hemolymph and could cause a fatal septicemia in the sea hare. Marine bacteria were usually cleared rapidly after injection. Marine bacteria were agglutinated by sea hare hemolymph, but *Serratia* was not. In addition, agglutinin titers fell dramatically after injection of marine bacteria but not after injection of *Serratia*. Therefore, it was suggested that *in vivo* marine bacteria interact with the lectin and are then phagocytized; since *S. marcescens* did not react with an agglutinin, it was not phagocytized.

Cells of the lesser octopus, *Eledone cirrosa*, did not phagocytize E in serum-free medium, but considerable uptake occurred in normal serum (Stuart, 1968). Human E opsonized with *Eledone* serum, washed, and treated with rabbit anti-*Eledone* serum will agglutinate. This indicated that some component of the octopus serum interacted with the E surface. Similar results were obtained with yeast and *E. coli*, suggesting the presence of an opsonin with affinity for a wide range of particles. No lectins are present in octopus blood. The major macromolecule in *Eledone* hemolymph is hemocyanin, which was suggested as a candidate for the opsonin.

Extensive work has been done with hemocytes of the crayfish *Parachaeraps bicarnatus*, which rarely ingested mammalian E *in vitro* but were capable of forming adhesion rosettes with E, under the proper conditions. The amount of adhesion of E with cell monolayers was dependent on opsonization with crayfish serum (McKay *et al.*, 1969). Crayfish serum adsorbed with mouse E lost its opsonizing capacity for these E; however, serum adsorbed with sheep E was still opsonic for mouse E. McKay and Jenkin (1970a) showed that the process of opsonization was rather specific; only crayfish or crab sera could stimulate E adhesion to crayfish hemocytes. Injection of bacterial vaccines produced increased protection against infection by the same microorganisms. However, no opsonins for bacteria and no bactericidal factors were detectable in normal or immune sera (McKay and Jenkin, 1970a). Increased phagocytic activity of hemocytes was suggested as the basis for the protective effect (McKay and Jenkin, 1970b). Immunization with bacterial vaccines did not influence the red cell opsonin titer.

So far there is little evidence to support the view that insects possess opsonins; the presence of hemagglutinins is well documented. Scott (1971b) described a non-γ-globulin lectin in the cockroach *Periplaneta americana* which was not inducible by injection of E. Sheep E exposed to the highest concentration of hemolymph which did not cause marked agglutination showed no more affinity for hemocytes than did untreated E. Another cockroach, *Blaberus craniifer*, possesses an agglutinin with specificity for certain erythrocytes, which is present in both late instar nymphs and adults (Anderson *et al.*, 1972). *Blaberus* hemolymph did not opsonize mammalian E. Hemocytes from this species could phagocytize and kill several kinds of bacteria *in vitro* (Anderson *et al.*, 1973a). Bacteria "opsonized" by pretreatment in hemolymph, or hemolymph concentrated by ultrafiltration, were not killed more efficiently than untreated bacteria. Indeed, thoroughly washed hemocyte monolayer preparations took up bacteria; therefore, hemolymph factors were not required for phagocytosis.

Naturally-occurring hemagglutinins are present in tunicates (Fuke and Sugai, 1972; Wright, 1974), including *Halocynthia pyriformis* (Anderson and Good, 1975). Fulton (1920) suggested that serum factors were required for carmine phagocytosis by *Ascidia atra* hemocytes. However, Anderson (unpubl.) was unable to show a similar opsonic effect during studies of E uptake by *Halocynthia* macrophages.

The role of complement (C) in phagocytosis and chemotaxis by vertebrate phagocytes is well known. Complement involvement in similar phenomena by invertebrate hemocytes has not as yet been assayed. Indeed it is unlikely that invertebrates possess the classical complement pathway; however, there is some indication some components of the alternate, i.e., properdin, pathway may be present in some invertebrates. This pathway utilizes the terminal components, C_3-9, and bypasses the early components and antibody. In mammals this pathway probably is the basis for the heat labile opsonin system, in which the initial complement components and antibody are not required for opsonization of Gram-negative bacteria. Patients with defects in the alternate complement pathway have marked susceptibility to infection, even if the early components and antibody levels are normal (Alper et al., 1969). Mammals deficient in the early components are frequently in good health.

Morgun (1950) first reported the presence of a factor similar to C_3 in invertebrate hemolymph. In mammals a factor (CVF) isolated from cobra venom by column chromatography can be shown to combine with a serum factor called C_3 proactivator (C_3PA) to activate the terminal complement components to cause the lysis of target erythrocytes. Lysis-inducing activity of CVF was described in the sera of Limulus, sipunculid worms, and starfish. Total hemolytic complement (indicated by the lysis of antibody-sensitized E) was not present in any of the invertebrate sera tested. A humoral factor functionally similar to C_3PA was described in cockroach hemolymph (Anderson et al., 1972). The formation of the hemolymph factor-CVF complex required divalent cations and was heat sensitive. The complex mediated the lysis of unsensitized sheep E via the terminal C components of frog serum. Neither CVF nor roach hemolymph alone could cause target cell lysis; lysis was shown to be complement dependent. Starfish hemolymph also contains a C_3PA-like factor (Day et al., 1972a). However, its molecular weight is estimated to be 2,000 (mammalian C_3PA is 80,000); such small size precludes enzymatic activity similar to C_3PA. The mode of action of invertebrate serum factor-CVF complexes on C activation has not as yet been defined. Nevertheless, Day et al. (1972b) reported that the interaction of starfish hemolymph factor, CVF, and purified C_3 generated anaphylatoxic reactions in guinea pig ileum preparations.

IV. CHEMOTAXIS OF INVERTEBRATE HEMOCYTES

The first quantitative study of chemotaxis by hemocytes from any invertebrate species appeared very recently (Schmid, 1975). Hemocytes from the snail Viviparus malleatus were centrifuged out of suspension into filters, of 8 μm pore size, which were

put into Boyden chambers. The hemocytes migrated through the
filters toward *Staphylococcus aureus* but only after the
bacteria interacted with a factor in the snail's hemolymph. This
factor was though to be an agglutinin; both bacterial and
mammalian E agglutinins were present in the hemolymph. No
chemotaxis was elicited by *S. aureus* in serum adsorbed with the
bacteria; such serum no longer agglutinated the bacteria. The
agglutination of both *S. aureus* and rabbit E was inhibited by
N-acetyl-D-glucosamine (GNAc). This indicated that both
agglutinins combine with GNAc, and it was suggested that combi-
nation was required to produce chemotactic agents. In this
light it was interesting that GNAc itself, after interaction
with hemolymph, was chemotactic.

This was another example of how certain invertebrate lectins
can serve as humoral recognition factors, active not only as
opsonins but in attracting phagocytes. The actual nature and
source of the above chemotactic effector molecules has not been
determined. Release of chemotactic factors from the lectin-
bacteria complexes was not shown in this study; it is unlikely
that the complexes themselves are chemotactic agents. It was
suggested that hemocytes might release chemotactic factors
after making initial random contact with the complexes. The
effect could be amplified by additional release of chemotactic
factors as more phagocytic hemocytes are attracted to the area.

V. BIOCHEMICAL EVENTS ASSOCIATED WITH PHAGOCYTOSIS

Recently, great progress has been made in the biochemical
characterization of mammalian phagocytic cell systems. The first
attempts to obtain similar data from invertebrate blood phagocytes were made by Anderson *et al.* (1973b) using cells of the
cockroach *Blaberus craniifer*. Similar findings have been subsequently reported for molluscan hemocytes; specifically those
of *Mercenaria mercenaria* (Cheng, 1975). In this section both
reports will be discussed and compared to what is known about
mammalian phagocytes under similar conditions.

Stahelin *et al.* (1956, 1957) first reported that phagocytosis
of tubercle bacilli by guinea pig polymorphonuclear leukocytes
(PMN) triggered a marked respiratory increment and increased
hexose monophosphate pathway (HMP) activity within the cells.
Soon thereafter, Sbarra and Karnovsky (1959) showed that
glycolytic inhibitors also inhibited particle ingestion.
Inhibition of cytochrome-linked respiration, Krebs cycle enzymes,
and oxidative phosphorylation did not alter ingestion or post-
phagocytic metabolic activity (Iyer *et al.*, 1971). Subsequently
genetically-controlled leukocyte metabolic disfuctions have been
correlated with impaired phagocytic function observed in certain
diseases (Holmes *et al.*, 1967; Baehner and Nathan, 1967;
Good *et al.*, 1968).

Metabolic information on *Blaberus* was obtained by studying cells actively phagocytizing latex particles or heat-killed *Staph. aureus* 502A.   Both particles were shown to be extensively ingested by direct microscopic observation and by spectrophotometric quantitation of latex after dioxane extraction of the hemocytes.   Similar data from *Mercenaria* hemocytes were obtained after phagocytic stimulation by heat-killed *Bacillus megaterium*.

A slight increment in O_2 utilization during phagocytosis was recorded for *Blaberus;* no increase was seen in the case of *Mercenaria*.   This increase was not comparable to that produced in mammalian PMN or casein-elicited peritoneal macrophages. Instead the results were similar to those seen with peritoneal macrophages obtained by saline lavage, cultured macrophages, and alveolar macrophages (Karnovsky *et al.*, 1970).

Particle ingestion stimulated glycogen and glucose catabolism and lactate production in both *Blaberus* and *Mercenaria* hemocytes. The effect of metabolic inhibitors, at concentrations not lethal to either cells or bacteria, on phagocytosis and killing of *Staph. aureus* 502A was studied in *Blaberus*.   The respiratory chain inhibitor KCN did not inhibit phagocytosis in either cell system; it also does not affect ingestion or killing by PMN. Inhibition of glycolysis by arsenate, iodoacetate, and fluoride resulted in decreased bactericidal capacity of cockroach phagocytes; this is also the case for mammalian PMN and macrophages.

Oxidative glucose metabolism by *Blaberus* hemocytes was measured by $^{14}CO_2$ quantitation during incubation with uniformly labeled ^{14}C-glucose. $^{14}CO_2$ production is a measure of both HMP and tricarboxylic acid cycle activity and was stimulated in *Blaberus* to a minimal extent during phagocytosis, in contrast to PMN.   Production of CO_2 from particular carbons of glucose can be assayed by using specifically-labeled ^{14}C-glucose. Increased CO_2 production from carbon 1 relative to carbon 6 reflects stimulated utilization of glucose by HMP.   C-1:C-6 ratio for PMN usually average about three, whereas the same ratio for macrophages are about one.   The ratio for phagocytizing *Blaberus* hemocytes is also approximately one, indicating little HMP activity.

The respiratory burst during PMN phagocytosis is mediated by flavoprotein oxidase resulting in the generation of H_2O_2 and oxidized nicotinamide adenine dinucleotide phosphate ($NADP^+$).   An increase in available $NADP^+$ stimulates the HMP.   Flavoprotein oxidase will reduce nitroblue tetrazolium (NBT) as described by Cagan and Karnovsky (1964).   Therefore, NBT reduction is associated with increased respiration and H_2O_2 production during phagocytosis by PMN.   NBT reduction cannot be detected in either insect or mollusk hemocytes.

The H_2O_2 generated, as described above, by PMN is antimicrobial in itself; it is also a component of the H_2O_2-myeloperoxidase (MPO)-halide antimicrobial system described by Klebanoff (1968). Efforts to detect the components of this system in *Blaberus* or *Mercenaria* hemocytes have been unsuccessful. When these cells were stained for MPO by the method of Kaplow (1965), only about 1% of the insect hemocytes contained MPO-positive granules, activity was not seen in mollusk phagocytes. Human PMN will iodinate ingested particles in concert with H_2O_2 and MPO. Quantitative iodination tests (method of Pincus and Klebanoff, 1971) using Na ^{125}I were negative for cockroach cells. We also looked for other antibacterial systems such as the cytoplasmic granule-associated cationic proteins described in mammalian cells by Zeya and Spitznagel (1966a, b). We were not able to isolate electrophoretically similar proteins from *Blaberus* lysosomes.

The above data indicate that the glycolytic pathway provides the metabolic energy for particle ingestion utilized by these invertebrate phagocytes. This is also the case for mammalian PMN and macrophages. Subsequent biochemical events typical of PMN are not found in the invertebrate phagocytes. The post-phagocytic activities of PMN include increased respiration, HMP activity, H_2O_2 generation and NBT reduction. Also, bactericidal mechanisms seen in PMN are absent in these invertebrate cells; these include the H_2O_2-MPO-halide system and cationic proteins. Most studies indicate that mammalian macrophages do not show a great respiratory increment during phagocytosis; they also show little HMP activity and atypical NBT reduction. Mammalian macrophages do not kill bacteria via the H_2O_2-MPO-halide system or by cationic proteins. Therefore, based on the available evidence, it would appear that invertebrate hemocytes share more characteristics with mammalian macrophages than with PMN, the principle circulating mammalian phagocytes.

VI. PHAGOCYTOSIS BY HEMOCYTE CELL LINES FROM ESTIGMENE ACREA

The hemocyte cell lines used in these studies were obtained from Dr. Robert A. Granados of the Boyce Thompson Institute, Yonkers, New York. Ten primary cell lines were prepared in in November, 1974. Hemolymph drops were obtained by cutting a proleg of surface-sterilized *E. acrea* larvae. Additional drops of hemolymph were added to the cultures at 2-day intervals until each 30 ml flask contained the cells from 3-4 drops of hemolymph. Cysteine (10^{-3}M) was added to inhibit melanization. The medium contained Grace's tissue culture medium, fetal calf serum and heat-inactivated *E. acrea* hemolymph (9:1:0.5). Suspended and attached cells (removed by trypsinization) were subcultured for the first time after 14 days, and subsequently 8 days later

using medium containing fresh, cell-free hemolymph (1.7%), as
well as heat-inactivated hemolymph. Only the two cultures
showing optimal growth were maintained and were designated
EA 1174 A and EA 1174 H. On the 7th subculture both lines were
growing well and two sublines, EA 1174 AS1 and EA 1174 HS1,
were adapted to a medium not containing fresh, cell-free
hemolymph. By July 1975, the cell lines and sublines had been
subcultured more than 40 times. The lines are presently
subcultured at 3-4 day intervals. The initial culture density
is 2-3 x 10^5 cells/ml, a density of 2-3 x 10^6 cells/ml is
usually attained within 72-96 hr. Both EA 1174 A and EA 1174 H
show similar growth rates with a doubling time of 20-21 hr.

These cell lines were developed to support the replication
of insect viruses, and have been shown to be susceptible to an
entomopoxvirus, a nuclear polyhedrosis virus, a cytoplasmic
polyhedrosis virus and an iridescent virus (Granados and
Naughton, 1975). This paper also gives details of the
establishment of the cell lines. The cell lines are composed of
plasmatocyte-type blood cells. These are phagocytic hemocytes
in vivo. The *in vitro* phagocytic potential of plasmatocytes
immediately after withdrawal from another insect species has
been studied by me in the past. Therefore, I was most interested in assaying the phagocytic ability of these cultured
hemocytes. The data presented here should be considered
preliminary but nevertheless are sufficient to describe several
characteristics of the behavior of these phagocytes toward a
variety of foreign particles. The effects of metabolic
inhibitors on particle ingestion are reported; statistical
analysis is not given because the studies are not complete.
However, the only effects discussed are based on relatively
great differences between experimental and control situations
which are highly reproducible and will, no doubt, prove to be
significant. At this time, few differences in phagocytic
reactions have been seen between the cell lines and sublines;
therefore, all statements can be assumed to be general, unless
otherwise indicated.

Phagocytosis of bacteria, erythrocytes, yeast, carmine, and
colloidal carbon was studied by microscopical examination of cell
monolayer preparations. The particles and suspended cultured
hemocytes were thoroughly mixed in small plastic tubes and
immediately drops of this mixture were placed on precleaned
coverslips. The coverslip preparations were incubated in
moist chambers at room temperature, about 23°C, for various
time intervals. The cells were then fixed with 100% methanol,
stained with hematoxylin and eosin, and mounted on slides.

Horse erythocytes adhered to about 50% of the hemocytes after
2 hr; adhesion was minimal (<5%) after 30 min. Actual ingestion
of horse E was rarely observed. However, formalized sheep red
cells formed rosettes around about 80% of the hemocytes by

30 min; 85-90% of the cells had adhered cells by 2 hr. The EA 1174 H (H) line seemed to be more active in E adhesion than the EA 1174 A (A) regardless of the presence of fresh hemolymph. Phagocytosis of the E was not extensive; less than 10% of the cells from either line contained ingested E after 24 hr.

Essentially similar results were seen when the cells were exposed to fresh or formalized yeast particles. Line A cells (20-40%) were covered with fresh or formalized yeast by 30 min after exposure. Little additional adhesion occurred during an additional 30 min. Slight uptake of formalized yeast by line A was observed; fresh yeast was not phagocytized. Similar adhesion and uptake of formalized yeast was seen with H cells; however, no adhesion or phagocytosis of unformalized yeast was detected with H cells.

Carmine particles were avidly removed from suspension by both A and H cells. Sixty to 90% of cells of both lines were coated with carmine after 30 min; essentially every cell had numerous particles adhering to its surface by 90 min. However, less than 5% of the cells contained ingested carmine 2 hr after introduction of the particles.

Several kinds of bacteria were added to the cultured plasmatocytes including *Serratia marcescens, Staphylococcus aureus* 502A, and *Escherichia coli* B. Little adhesion and no phagocytosis of *Serratia marcescens* was seen. In the case of *Staph. aureus,* after 1 hr, about 90% of the A cells and 70% of the H cells had adhering bacteria; however, neither line ingested more than 10 bacteria/100 cells, even after 6 hr. *E. coli* were attached to about 90% of A cells and 50% of H cells after 1 hr, but little or no uptake occurred in either line.

We attempted to assay cellular bactericidal capacity of A and H cells incubated with test bacteria using the technique of Anderson *et al.*, (1973a). The cells were washed several times and resuspended in antibiotic-free Grace's tissue fluid and bacteria (the species used in the monolayer assays described above) added; aliquots were diluted and plated to quantify the numbers of viable bacteria remaining at various time intervals. As was expected on the basis of the coverslip studies, the cells did not have any significant bactericidal capacity. The cells were viable after incubation with the bacteria, on the basis of trypan blue exclusion. Any killing of bacteria by the hemocyte suspensions was attributed to the presence traces of the antibiotic gentamicin (Schering Division, Diagnostic Products) which was added to the subcultures, and was apparently not completely removed when the cells were washed. Cells from gentamicin-free cultures were not bactericidal.

A colloidal suspension of carbon particles were obtained from India ink by centrifugation (the least dense layer was used), washed, and resuspended in Grace's medium. The percentage of A or H having adhering carbon particles was seen to increase with

time until at least 90% were coated. This reaction was rapid
with most cells covered with carbon particles by 30 min.
In addition to adhesion, marked phagocytosis of carbon could
be seen. There was a steady increase in the numbers of cells
containing ingested carbon until about 80% contained particles
by 6 hr. If the cells and carbon were incubated at $1°C$, little
adhesion and no phagocytosis was observed.

Ingestion of carbon was used as a measure of phagocytosis to
study the effects of antimetabolites on the cultured hemocytes.
In these experiments trypan blue exclusion tests were always
performed to determine the viability of the cells. Inhibitor
concentrations were adjusted to exert their metabolic effects
without killing the cells. The effects of inhibitors of
glycolysis and the electron transport system on phagocytosis
were studied.

Cyanide inhibits the cytochrome $a + a_3$ complex (cytochrome
oxidase) which mediates the final transfer of electrons to
molecular oxygen. Cyanide concentrations of 10^{-3} M and less
caused only minimal cell death (<10%), even after 6 hr. Concentrations lower than 10^{-5} did not affect phagocytosis; 10^{-4}
M was slightly inhibitory. A higher concentration (10^{-3} M) caused
marked inhibition of carbon uptake at all time intervals studied
(1, 4, and 6 hr). Phagocytosis by insect, mollusk, and mammalian
PMN hemocytes was not affected by similar (8×10^{-4} M) cyanide
coccentrations (Anderson et $al.$, 1973b; Cheng, 1975; Cohn and
Morse, 1960). In light of this difference we are examining the
cyanide effect in more detail.

Glycolysis supplies the energy requirement for particle
ingestion by both vertebrate and invertebrate phagocytic blood
cells in short-term culture. Therefore, the effect of three
glycolytic inhibitors on carbon phagocytosis by these established
cell lines was studied. Iodoacetate is an alkylating agent
which inhibits glyceraldehyde 3-phosphate dehydrogenase. Arsenate
also inhibits glycolysis at the same point, uncoupling oxidation
and phosphorylation in the triose oxidation step. The second
reaction of the glycolytic sequence in which a high-energy
phosphate bond is generated (the conversion of 2 phosphoglycerate
to phosphoenolpyruvate catalyzed by enolase) is strongly inhibited by fluoride, particularly in the presence of phosphate.

Iodoacetate ($\leq 10^{-3}$ M) is not toxic to either A or H cells,
but will inhibit carbon uptake at concentrations as low as
10^{-5} M. Arsenate was toxic to both cell lines at levels above
10^{-4} M, particularly after several hours of incubation. Nontoxic
arsenate levels (10^{-5} M) cause substantial phagocytic inhibition.
Incubation of A or H cells in 10^{-3} M, or less, sodium fluoride
for at least 4 hr caused little or no cell death; however,
$\geq 10^{-5}$ M usually inhibited carbon phagocytosis. Therefore, it
would appear that these cultured insect hemocytes also use
glycolysis to provide energy for phagocytosis.

VII. ACKNOWLEDGMENTS

I would like to express my appreciation to Dr. Robert R. Granados of the Boyce Thompson Institute for making the *Estigmene* hemocyte lines available to me and for his helpful comments. Mrs. Marybeth Naughton maintained the cell lines and kept me constantly supplied with experimental material.

I am grateful to Miss Molly Cook and Mrs. Lois A. Jordan for their technical assistance.

This was supported by a grant from the Whitehall Foundation and Grant CA-08748 from the National Cancer Institute.

SUMMARY

This paper summarizes information currently available on recognition and phagocytosis of foreign particles by invertebrate blood cells held in short-term culture. New observations of the phagocytic potential of recently established insect hemocyte lines are also presented.

Phagocytosis by hemocytes of many invertebrate species has been studied *in vitro*, and has been documented on film, photomicrographs, and electromicrographs. All foreign particles are not ingested with equal avidity by macrophages from any given species. In many cases, chemical alteraction of the particle's surface by gluteraldehyde, formaldehyde, etc. will enhance its ingestion by phagocytes. The phagocytic activity of crustacean hemocytes can be stimulated as a result of the injection of bacterial vaccines. This type of activation is probably non-specific in that the phagocytosis of all kinds of particles is augmented.

There is good evidence that certain invertebrate species have humoral factors which promote phagocytosis by opsonizing foreign particles. However, in many cases, these factors are not absolutely required for phagocytosis. Hemagglutinins may serve as opsonins, but certainly all invertebrate opsonins are not hemagglutinins and vice versa. Classical mammalian antibodies have never been found in any invertebrate, and there is no indication that metazoan invertebrate phagocytes have immunoglobulin receptors. Acellular forms (protozoans and slime molds) do phagocytize antibody-sensitized particles more readily than untreated particles; the physiological significance of this is unclear. In mammals complement plays an important role in phagocytosis. There is some indication that invertebrates may possess the terminal complement components (C_3-C_9), but direct evidence for complement involvement in phagocytosis in these animals is lacking.

The first quantitative studies of chemotaxis by invertebrate hemocytes have been performed. Apparently, a hemolymph factor,

probably an agglutinin, reacts with bacteria to form a chemotactic complex. Release of chemotactic factors from this complex
was not shown; possibly cells making random contact with the
complexes release chemotactic factors after the bacteria are
ingested.

The results of comparative metabolic studies of phagocytizing
insect hemocytes and mollusk hemocytes show many similarities.
The involvement of the glycolytic pathway in particle uptake
has been shown by direct chemical analysis and by the use of
enzyme inhibitors. Glycolysis provides the energy for phagocytic
ingestion in all animal species so far studied. The post-phagocytic
events recorded in mammalian polymorphonuclear cells include
marked oxygen utilization, hexose monophosphate pathway activation, H_2O_2 production, and nitroblue tetrazolium reduction.
These phenomena are not seen in invertebrate phagocytes and in
most mammalian macrophages. Polymorphonuclear cells kill bacteria
via the myeloperoxidase - H_2O_2 - halide system and lysosomal
cationic proteins. These mechanisms are lacking in invertebrate
cells and mammalian macrophages. These similarities suggest that
invertebrate phagocytic hemocytes may be properly referred to as
macrophages.

Preliminary results from the first studies of the phagocytic
behavior and metabolism of established insect hemocyte cell
lines are given in this paper. Formalized sheep erythrocytes
adhered quickly and in great numbers, forming rosettes around
the hemocytes. Actual red cell ingestion was infrequent. Other
kinds of particles such as yeast, bacteria, and carmine also
readily adhered but were seldom phagocytized. The cell lines
had little or no bactericidal activity against *Staphylococcus
aureus* 502A, *Serratia marcescens*, or *Escherichia coli*. The
cells were able to ingest colloidal carbon and about 80% contained particles after 6 hr. Uptake of carbon was used to
measure phagocytosis to study the effects of various antimetabolites. The enzyme inhibitors selectively inhibited steps
in the glycolytic pathway or electron transport without killing
the cells on the basis of trypan blue exclusion. Unlike
circulating hemocytes, phagocytosis by these cells was somewhat
inhibited by cyanide. Glycolytic inhibitors, i.e., iodoacetate,
arsenate, and fluoride, substantially inhibited phagocytosis of
carbon.

REFERENCES

Alper, C. A., Abramson, N., Johnston, R. B., Jr., Jandl, H.,
    and Rosen, F. S.(1969). Increased susceptibility to
    infection associated with abnormalities of complement-
    mediated functions and of the third component of complement
    (C3). *N. Engl. J. Med.*, 282. 350-354.

Anderson, R. S. (1975). Phagocytosis by invertebrate cells
in vitro: Biochemical events and other characteristics
compared with vertebrate phagocytic systems. *In
"Invertebrate Immunity,"* (K. Maramorsch and R. E. Shope,
eds.), p. 365, Academic Press, New York.

Anderson, R. S., Day, N. K. B., and Good, R. A. (1972). Specific
hemagglutinin and a modulator of complement in cockroach
hemolymph. *Infect. Immun.,* 5, 55-59.

Anderson, R. S., Holmes, B., and Good, R. A. (1973a). *In vitro*
bactericidal capacity of *Blaberus craniifer* hemocytes.
J. Invertebr. Pathol., 22, 127-135.

Anderson, R. S., Holmes, B., and Good, R. A. (1973b). Comparative
biochemistry of phagocytizing insect hemocytes. *Comp.
Biochem. Physiol.,* 46B, 595-603.

Anderson, R. S., and Good, R. A. (1975). Naturally occurring
hemagglutinin in a tunicate *Halocynthia pyriformis. Biol.
Bull.,* 148, 357-369.

Anderson, R. S., and Good, R. A. (1976). Opsonic involvement
in phagocytosis by mollusk hemocytes. *J. Invertebr. Pathol.*
27, 57-64.

Baehner, R. L., and Nathan, D. G. (1967). Leukocyte oxidase:
Defective activity in chronic granulomatous disease.
Science, 155, 835-836.

Bang, F. B. (1961). Reaction to injury in the oyster (*Crassostrea
virginica*). *Biol. Bull.,* 121, 57-68.

Cagen, R. H., and Karnovsky, M. C. (1964). Enzymatic basis of
the respiratory stimulation during phagocytosis. *Nature*
204, 255-257.

Cheng, T. C. (1976). Aspects of substrate utilization and energy
requirement during molluscan phagocytosis. *J. Invertebr.
Pathol.,* 27, 263-268.

Cohn, Z. A., and Morse, S. I. (1960). Functional and metabolic
properties of polymorphonuclear leucocytes. I. Observations
on the requirements and consequences of particle ingestion.
J. Exp. Med., 111, 667-687.

Day, N., Geiger, H., Finstad, J., and Good, R. A. (1972a). A
starfish hemolymph factor which activates vertebrate
complement in the presence of cobra venom factor. *J. Immunol.,*
109, 164-167.

Day, N. K. B., Good, R. A., and Muller-Eberhard, H. J. (1972b).
"C3 Proactivator" in starfish hemolymph. *Fed. Proc.,*
31, 788 Abstr.

Fuke, M. T., and Sugai, T. (1972). Studies on the naturally
occurring hemagglutinin in the coelomic fluid of an Ascidian.
Biol. Bull., 143, 140-149.

Fulton, J. F., Jr. (1920). The blood of *Ascidia atra. Acta Zool.,*
1, 381-432.

Gerisch, G., Luderitz, O., and Ruschmann, E. (1967). Antikorper
forden die phagozytose von bakterien durch amoben.
Z. Naturforsch (B), 22, 109.

Githens, S., III, and Karnovsky, M. L. (1973). Phagocytosis by
   the cellular slime mold *Polysphondylium pallidum* during
   growth and development. *J. Cell Biol.*, 58, 536-548.
Good, R. A., Quie, P. G., Windhorst, D. B., Page, A. R.,
   Rodey, G. E., White, J. E., and Holmes, B.(1968). Fatal
   (chronic) granulomatous disease of childhood: A hereditary
   defect of leukocyte function. *Semin. Hamatol.*, 5, 215-254.
Goodall, R. J., and Thompson, J. E. (1971). A scanning electron
   microscopic study of phagocytosis. *Exp. Cell Res.*,
   64, 1-8.
Grace, T. D. C. (1962). The development of a cytoplasmic
   polyhedrosis in insect cells grown *in vitro*. *Virology*,
   18, 33-42.
Grace, T. D. C. (1971). *In "Invertebrate Tissue Culture,"* Vol. I.
   (C. Vago, ed.) Academic Press, New York.
Grace, T. D. C., and Day, M. F. (1963). Film, C.S.I.R.O.
Granados, R. R., and Naughton, M. (1975). Replication of
   Amsacta moorei entomopoxvirus and *Autographa californica*
   nuclear polyhedrosis virus in hemocyte cell lines from
   Estigmene acrea. *In "Proc. 4th Int. Conf. Invertebr.*
   Tissue Culture." (K. Maramorosch and E. Kurstak, ed.)
   Academic Press, New York (in press).
Holmes, B., Page, A. R., and Good, R. A. (1967). Studies of the
   metabolic activity of leukocytes from patients with a genetic
   abnormality of phagocytic function. *J. Clin. Invest.*,
   46, 1422-1431.
Iyer, G. Y. N., Islam, M. F., and Quastel, J. M. (1961).
   Biochemical aspects of phagocytosis. *Nature*, 192, 535-541.
Johnson, P. T. (1969). The coelomic elements of sea urchins
   (*Strongylocentrotus*). III. *In vitro* reactions to bacteria.
   J. Invertebr. Pathol., 13, 42-62.
Johnson, P. T., Chien, P. K., and Chapman, F. A. (1970). The
   coelomic elements of sea urchins (*Strongylocentrotus*).
   V. Ultrastructure of leukocytes exposed to bacteria.
   J. Invertebr. Pathol., 16, 466-469.
Kaplow, L. S. (1965). Simplified myeloperoxidase stain using
   benzidene dihydrochloride. *Blood*, 26, 214-219.
Karnovsky, M. L., Simmons, S., Glass, E. A., Shafer, A. W.,
   and D'arcy Hart, P. (1970). Metabolism of macrophages.
   In "Mononuclear Phagocytes." (R. van Furth, ed.) pp.
   103-117. F. A. Davis, Philadelphia.
Klebanoff, S. J. (1968). Myeloperoxidase-halide-hydrogen peroxide
   antimicrobial system. *J. Bacteriol.*, 95, 2131-2138.
McKay, D., and Jenkin, C. R. (1970a). Immunity in the invertebrates. The role of serum factors in phagocytosis of
   erythrocytes by haemocytes of the fresh water crayfish
   (*Parachaeraps bicarinatus*). *Aust. J. Exp. Biol. Med. Sci.*,
   48, 139-150.

McKay, D., and Jenkin, C. R. (1970b). Immunity in the inverte-brates. Correlation of the phagocytic activity of haemocytes with resistance to infection in the crayfish *(Parachaeraps bicarinatus)*. *Aust. J. Exp. Biol. Med. Sci.*, 48, 609-617.

McKay, D., and Jenkin, C. R. (1970c). Immunity in the inverte-brates. The fate and distribution of bacteria in normal and immunized crayfish *(Parachaeraps bicarinatus)*. *Aust. J. Exp. Biol. Med. Sci.*, 48, 599-607.

McKay, D., Jenkin, C. R., and Rowley, D. (1969). Immunity in the invertebrates. I. Studies on the naturally occurring haemagglutinins in the fluid of invertebrates. *Aust. J. Exp. Biol. Med. Sci.*, 47, 125-134.

Morgun, G. I. (1950). Complement in invertebrates. *Mikrobiol. Zh.* (Kiev), 11, 43-50.

Pauley, G. B., Granger, G. A., and Krassner, S. M. (1971a). Characterization of a natural agglutinin in the hemolymph of the california sea hare, *Aplysia californica*. *J. Invertebr. Pathol.*, 18, 207-218.

Pauley, G. B., Krassner, S. M., and Chapman, F. A. (1971b). Bacterial clearance in the california sea hare, *Aplysia californica*. *J. Invertebr. Pathol.*, 18, 227-239.

Pincus, S. H., and Klebanoff, S. J. (1971). Quantitative leukocyte iodination. *N. Engl. J. Med.*, 284, 744-750.

Prowse, R. H., and Tait, N. N. (1969). *In vitro* phagocytosis by amoebocytes from the haemolymph of *Helix aspersa* (Muller). I. Evidence for opsonic factor(s) in serum. *Immunology*, 17, 437-443.

Rabinovitch, M., and DeStefano, M. (1970). Interactions of red cells with phagocytes of the wax-moth *(Galleria mellonella, L.)* and mouse. *Exp. Cell Res.*, 59, 272-282.

Rabinovitch, M., and DeStefano, M. (1971a). Phagocytosis of erythrocytes by *Acanthamoeba* sp. *Exp. Cell Res.*, 64, 275-284.

Rabinovitch, M., and DeStefano, M. (1971b). Antibody and plant agglutinins stimulate phagocytosis of erythrocytes by *Acanthamoeba*. *Nature* 234, 414-415.

Ratcliffe, N. A., and Rowley, A. F. (1974). *In vitro* phagocytosis of bacteria by insect blood cells. *Nature*, 252, 391-392.

Sbarra, A. J., and Karnovsky, M. L. (1959). The biochemical basis of phagocytosis. I. Metabolic changes during the ingestion of particles by polymorphonuclear leukocytes. *J. Biol. Chem.* 234, 1355-1362.

Schmid, L. A. (1975). Chemotaxis of hemocytes from the snail *Viviparus malleatus*. *J. Invertebr. Pathol.*, 25, 125-131.

Scott, M. T. (1971a). Recognitition of foreignness in inverte-brates. II. *In vitro* studies of cockroach phagocytic haemocytes. *Immunology*, 21, 817-828.

Scott, M. T. (1971b). A naturally occurring hemagglutinin in the
    hemolymph of the american cockroach. *Arch. Zool. Exp.
    Gen.*, 112, 73–80.
Stahelin, H., Karnovsky, M. L., Farnham, A. E., and Suter, E.
    (1957). Studies on the interaction between phagocytes and
    tubercle bacilli. III. Some metabolic effects in guinea
    pigs associated with infection with tubercle bacilli.
    J. Exp. Med., 105, 265–277.
Stahelin, H., Suter, E., and Karnovsky, M. L. (1956). Studies
    on the interaction between phagocytes and tubercle bacilli.
    I. Observations on the metabolism of guinea pig leucocytes
    and the influence of phagocytosis. *J. Exp. Med.*, 104,
    121–136.
Stuart, A. E. (1968). The reticulo-endothelial apparatus of the
    lesser octopus, *Eledone cirrosa*. *J. Pathol. Bacteriol.*,
    96, 401–402.
Tripp, M. R. (1966). Hemagglutinin in the blood of the oyster
    Crassostrea virginica. *J. Invertebr. Pathol.*, 8, 478–
    484.
Tripp, M. R., and Kent, V. E. (1967). Studies on oyster cellular
    immunity. *In Vitro*, 3, 129–135.
Vago, C. (1964). "Culture de tissus d'invertebres." Service du
    Film du Recherche Scientifique, Paris.
Weisman, R. A., and Korn, E. D. 1967 Phagocytosis of latex
    beads by *Acanthamoeba*. I. Biochemical properties.
    Biochemistry, 6, 485–497.
Wright, R. K. (1974). Protochordate immunity. I. Primary immune
    response of the tunicate *Ciona intestinalis* to vertebrate
    erythrocytes. *J. Invertebr. Pathol.*, 24, 29–36.
Zeya, H. I., and Spitznagel, J. K. (1966a). Cationic proteins of
    polymorphonuclear leukocyte lysosomes. I. Resolution of
    antibacterial and enzymatic activities. *J. Bacteriol.*,
    91, 750–754.
Zeya, H. I., and Spitznagel, J. K. (1966b). Cationic proteins of
    polymorphonuclear leukocyte lysosomes. II. Composition,
    properties, and mechanism of antibacterial action. *J.
    Bacteriol.*, 91, 755–762.

Biochemical and Ultrastructural Evidence for the Double Role of Phagocytosis in Molluscs: Defense and Nutrition

Thomas C. Cheng

INSTITUTE FOR PATHOBIOLOGY
CENTER FOR HEALTH SCIENCES
LEHIGH UNIVERSITY
BETHLEHEM, PENNSYLVANIA

I. INTRODUCTION

Haeckel (1862) was the pioneer in reporting that the hemolymph cells of molluscs are capable of phagocytosis. Since that initial report, a number of individuals have demonstrated the phagocytosis of nonself materials by molluscan cells (see Feng, 1967; Cheng, 1967; for reviews).

The role of hemolymph cells in internal defense originated with Stauber's (1950) study which involved tracing the fate of India ink experimentally injected intracardially into the American oyster, *Crassostrea virginica*. In brief, he reported that the carbon particles are eventually removed from oysters as a result of ink-laden phagocytes migrating to the exterior through the epithelial surfaces of the alimentary tract, digestive diverticula, palps, mantle, and pericardium. It should be borne in mind that carbon particles are essentially inert to enzymatic degradation.

Stauber's study was followed by those of Tripp (1958a, b, 1960) who injected a variety of foreign materials, including digestible and nondigestible particulate materials, into *C. virginica*. It was discovered as a result that all became phagocytized; however, Tripp was able to demonstrate that the digestible particles, i.e., rabbit erythrocytes, vegetative bacterial cells, and some yeast cells, were eliminated via the migration of particle-laden phagocytes across epithelial borders as well as by intracellular digestion within phagocytes. That intracellular digestion of nonself materials by molluscan phagocytes does occur has been verified by Feng (1959, 1965) who demonstrated that soluble materials injected into *C. virginica* are pinocytized by phagocytes followed by intracellular digestion.

As a result of the pioneering studies cited above and several subsequent ones (Reade, 1968; Arcadi, 1968; Stuart, 1968; Cheng *et al.*, 1969; Bayne and Kime, 1970; Pauley and Krassner, 1972; Sminia, 1972; Bayne, 1973), the principle is now well established that hemolymph cells of molluscs will readily phagocytize foreign materials introduced into their bodies. Furthermore, as a rule, such materials, if digestible, are degraded by intracellular digestion and some are removed when cells containing such materials migrate to the exterior through epithelial borders. Thus phagocytosis has been established as one of the main internal defense mechanisms in molluscs. The other mechanisms are encapsulation and nacrezation (see Cheng and Rifkin, 1970).

Molluscan hemolymph cells are involved in functions other than defense. For example, they have been reported to be involved in wound repair (Pauley and Sparks, 1965; Des Voigne and Sparks, 1968; Pauley and Heaton, 1969), shell repair (Wagge, 1951, 1955), and digestion (Yonge, 1923, 1926; Takatsuki, 1934; Yonge and Nicholas, 1940; Zacks and Welsh, 1953; Zacks, 1955). The question, however, that had not been raised was: Is there a correlation between the immunologic and digestive functions of molluscan hemolymph cells?

This question was raised in our laboratory as a result of the finding by Cheng and Cali (1974) and Cheng *et al.* (1974) that glycogen granules are synthesized and temporarily stored in the secondary phagosomes of *C. virginica* subsequent to the phagocytosis of bacteria. Furthermore, it had been found that in time the glycogen is discharged into serum and presumably eventually employed as an energy source (Cheng and Cali, 1974; Cheng, 1975).

Since the earlier studies involved tracing the fate of phagocytized bacteria by employing electron microscopy, it was decided to verify our earlier interpretations by tracing the fate of radioactively labelled bacteria.

II. MATERIALS AND METHODS

A. *MAINTENANCE OF OYSTERS*

The specimens of the American oyster, *C. virginica*, used in this study were obtained from the coast of Connecticut through the courtesy of Dr. S. Y. Feng. They were maintained in recirculating seawater tanks at 22°C with a salinity of 20°/00. Each of the oysters used measured between 8 and 10 cm in length and weighed between 140 and 160 g. Those injected with radioactive bacteria, those receiving sham injections, and those not injected were held in separate aerated aquaria each containing 30 liters of sea water.

B. *BACTERIAL INOCULA*

Five hundred ml of tryptose phosphate broth spiked with 250 µCi of ^{14}C-labelled D-(U-^{14}C) glucose (ICN Corporation, Irvine, California) were employed to culture *Bacillus megaterium*. After 24 hr of culture, the bacteria were heat killed, washed five times in sterile deionized water, and resuspended in Millipore-filtered sea water with a salinity of 20°/00.

A 0.5 ml inoculum containing 5.01×10^8 bacteria was injected through holes previously drilled on the left valve of each of 12 experimental oysters in the region overlaying the heart. After the bacterial suspension had been introduced, the holes were sealed with cotton gauze and melted paraffin.

Sterile sea water was similarly injected into each of three sham-treated oysters. Three additional oysters not injected with any material also served as controls.

C. *EXTRACTION PROCEDURES*

One valve of each of three oysters that had been injected with ^{14}C-labelled *B. megaterium* was removed at 6, 16, 24, and 72 hr post-injection and from each a sample of hemolymph was collected from the mantle cavity and pericardial region by employing sterile

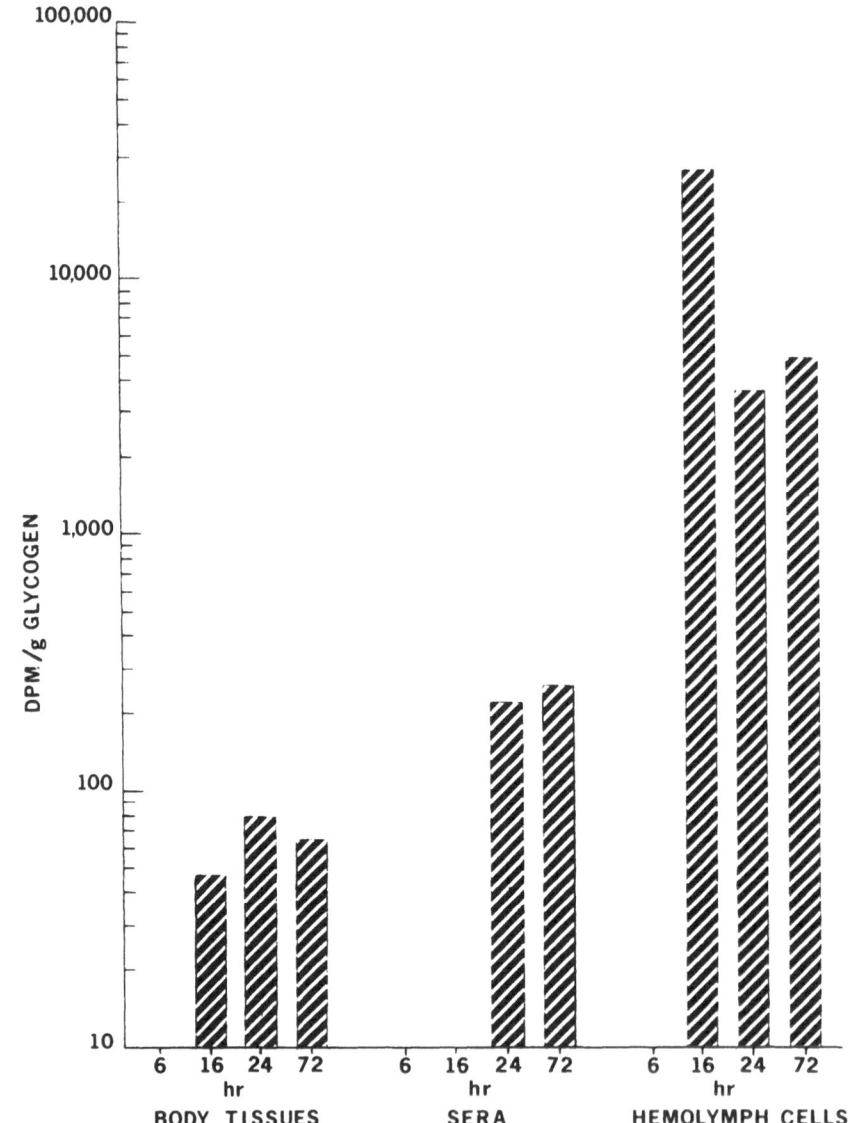

Fig. 1.   Disintegrations per minute (DPM) per gram of glycogen
          from the body tissues, sera, and hemolymph cells of
          Crassostrea virginica extracted at 6, 16, 24, and 72 hr
          after injection of ^{14}C-labelled *Bacillus megaterium*.

syringe and needle. Hemolymph samples were similarly collected
from the sham-injected oysters at 72 hr post-injection and from
untreated ones at the same time. In all cases the soft tissues
were subsequently removed and perfused in running water for 30
min.

The cells in all of the hemolymph samples were concentrated by
centrifugation at 5000 g and washed three times in sterile sea
water.

The hemolymph cells, sera, and soft tissues were maintained
frozen until employed. None of the samples were thus stored for
more than 2 weeks.

By employing a modification of the method of Scott (1969),
mono- and disaccharides were selectively precipitated by forming
the corresponding phenylosazones, which are insoluble in water.
Specifically, 2 ml of a 0.5% solution of 2,4-dinitrophenylhydra-
zene in 2 N HCl was added to 1 ml of serum and the mixture was
permitted to react for 30 min in a boiling water bath. The resulting crystals were filtered, washed with water, and the filter
papers on which the crystals collected were cut into pieces and
the 2,4-dinitrophenylosazones were dissolved in 0.15 ml of
dioxane in scintillation vials.

Glycogen was extracted from hemolymph cells, sera, and body
tissues by employing a modification of the method of Polyglase
et al. (1952). Specifically, each sample was digested with 30%
KOH for 2 hr in a boiling water bath followed by clarification,
if necessary, by filtering through glass wool. Subsequently, a
precipitate was formed by adding 1.2 volumes of 95% ethanol.
After centrifugation, the crude glycogen again was treated with
30% KOH for 2 hr and reprecipitated with 1.2 volumes of 95%
ethanol. Subsequently, the glycogen was washed four times with
52% ethanol, dissolved in water, and centrifuged at 27,138 g to
remove any particulate impurities. Finally, 1.2 volumes of 95%
ethanol was employed to reprecipitate the glycogen, which was
then washed four times with 52% ethanol followed by drying.

D. MEASUREMENT OF ^{14}C ACTIVITY

Osazone crystals were prepared for liquid scintillation
counting by adding 10 ml of LSC Cocktail Formulation III (J. T.
Baker Chem. Co., Phillipsburg, N. J.) to vials each of which
contained 0.15 ml of dioxane-dissolved crystals.

The samples of glycogen extracted from sera and body tissues
were prepared by dissolving 10 and 50 mg of glycogen,
respectively, in 0.1 ml of water in vials and adding 10 ml of
scintillation cocktail. The hemolymph cell glycogen samples were
prepared by dissolving approximately 1 mg of glycogen in 0.1 ml
of water and adding 10 ml of the cocktail. All counts were made
with a Packard Tri-Carb Model 3330 liquid scintillation spectro-
meter with a window setting of 50-1000 and a gain of 6%.

III. RESULTS

A. *COUNTING EFFICIENCY*

The counting efficiency was determined for each group of samples by spiking with ^{14}C-labelled toluene. The efficiencies were determined to be 68.2±4% for the osazone samples, 72.6±4% for the hemolymph cell glycogen samples, 70.1±4% for the serum glycogen samples, and 71.4±4% for the body tissue glycogen samples.

B. *RADIOACTIVITIES IN TISSUES*

None of the 12 experimental oysters had detectable levels of ^{14}C in the osazone preparations at the specified time intervals post-injection; however, ^{14}C activity was detected in the glycogen extracted from the hemolymph cells, sera, and body tissues (Fig. 1).

Disintegrations/min/g of glycogen were calculated and plotted against the times at which the samples were examined. No ^{14}C activity was detected in samples of glycogen at 6 hr post-injection (Fig. 1); however, activity was detected associated with the glycogen extracted from both the body tissues and hemolymph cells at 16 hr post-injection (Fig. 1). Furthermore, 8 hr after ^{14}C activity was first detected in these samples, activity was detected in the glycogen extracted from sera, i.e., at 24 hr post-injection. Furthermore, each glycogen sample revealed activity through the 72nd hr post-injection (Fig. 1).

The radioactivity associated with the glycogen extracted from sera was approximately 3.5 times that of each gram of glycogen extracted from the body tissues,and the glycogen extracted from hemolymph cells was 100 and 20 times those of the serum glycogen at 16 hr and 24 hr, respectively.

None of the glycogen samples extracted from the hemolymph cells, sera, and body tissues of the sham-injected and uninjected oysters included radioactivity.

IV. DISCUSSION AND CONCLUSIONS

The electron microscope studies by Cheng and Cali (1974), Cheng (1975), and to some extent by Cheng *et al.* (1974) have provided evidence that in the American oyster, *C. virginica*, phagocytized bacteria are initially enclosed within primary phagosomes where enzymatic degradation is initiated. This is followed by the transfer of some of the bacterial constituents to secondary phagosomes. Other constituents of the phagocytized bacteria, such as lipoidal and proteinaceous materials, are discharged from the phagocyte, primarily granulocytes, when the primary phagosome makes contact with the surface membrane and communicates with the exterior through an aperture (Cheng, 1975).

These morphological evidences confirm and extend the previously available information that molluscan hemolymph cells are capable of degrading phagocytized nonself materials.

In addition, the electron microscopical evidence supporting the interpretation that the carbohydrate constituents of the phagocytized bacteria are by some yet undetermined mechanism converted to glycogen within secondary phagosomes. Furthermore, the eventual discharge of this polysaccharide into serum suggests that there is a direct correlation between cellular immunity, i.e., phagocytosis and intracellular degradation, and carbohydrate nutrition.

Relative to nutrition, it is noted that Bayne (1973), working with the terrestrial gastropod *Helix pomatia,* has observed that after injection, bacterial ^{14}C is eventually distributed increasingly more evenly throughout all of the tissues of this snail. This finding, plus our earlier more detailed interpretation of electron microscopical findings (Cheng and Cali, 1974; Cheng, 1975), especially the latter, are supported by the demonstration that there is an increase in the amount of glycogen in the serum following phagocytosis of bacteria.

The data reported herein demonstrate that glycogen synthesis from bacterial constitutents in phagocytes is first detectable after 6 hr post-injection and reaches a peak soon thereafter, i.e., at 24 hr in the case of body tissues, 72 hr in the case of sera, and 16 hr in the case of hemolymph cells (Fig. 1). Glycogen, however, is still released from the phagocytes after 16 hr post-injection. This is indicated by the simultaneous drop in radioactivity associated with the glycogen of hemolymph cell origin from 16 to 24 hr post-injection and the initial appearance at 24 hr post-injection of radioactivity in sera associated with the glycogen of bacterial origin (Fig. 1). It is noted that the latter activity is further increased at 72 hr post-injection. Furthermore, activity is first detected in the glycogen extracted from the body tissues at 16 hr post-injection, which coincides with the initial detection of radioactivity associated with the glycogen extracted from the hemolymph cells. This phenomenon is to be expected since hemolymph cells probably are trapped in the mantle and mucus of the oyster (Cheng, 1975).

It is our opinion that Bayne's (1973) results imply that the ^{14}C in the serum glycogen is eventually utilized by all of the tissues in the mollusc's body; however, the mechanism for its transport and uptake remains to be elucidated. Presumably, the glycogen is initially degraded to glucose, which could diffuse into the various body tissues and be utilized.

The fact that no activity was detected in the serum osazone preparations indicates that the ratio of glucose of bacterial origin to preexisting glucose in *C. virginica* was not sufficiently high to enable detection under the conditions of the experiment.

V. ACKNOWLEDGMENT

The technical assistance of Mr. Barney Rudo is gratefully acknowledged.

SUMMARY

Specimens of *Crassostrea virginica* were injected with ^{14}C-labeled *Bacillus megaterium,* and glycogen was extracted from their hemolymph cells, sera, and body tissues at 6, 16, 24, and 72 hr post-injection.
^{14}C was first detected in glycogen extracted from the hemolymph cells and body tissues at 16 hr post-injection and at 24 post-injection in the case of serum glycogen.
The data presented support earlier observations that the degradation of phagocytized bacteria in molluscan hemolymph cells leads to the synthesis of glycogen from sugar of bacterial origin and its eventual release from phagocytes.

REFERENCES

Arcadi, J. A. (1968). Tissue response to the injection of charcoal into the pulmonate gastropod *Lehmania poirieri.* *J. Invertebr. Pathol.,* 11, 59-62.
Bayne, C. J. (1973). Molluscan internal defense mechanism: the fate of ^{14}C-labeled bacteria in the land snail *Helix pomatia* (L.). *J. Comp. Physiol.,* 86, 17-25.
Bayne, C. J. and Kime, J. B. (1970). *In vivo* removal of bacteria from the hemolymph of land snail *Helix pomatia* (Pulmonata: Stylommatophora). *Malacol. Rev.,* 3, 103-113.
Cheng, T. C. (1967). Marine molluscs as hosts for symbioses. *Adv. Mar. Biol.,* 5, 1-424.
Cheng, T. C. (1975). Functional morphology and biochemistry of molluscan phagocytes. *Ann. N.Y. Acad. Sci.,* 266, 343-379.
Cheng, T. C. and Cali, A. (1974). An electron microscope study of the fate of bacteria phagocytized by granulocytes of *Crassostrea virginica.* *Contemp. Top. Immunobiol.,* 4, 25-35.
Cheng, T. C. and Rifkin, E. (1970). Cellular reactions in marine molluscs in response to helminth parasitism. In "Diseases of Fish and Shellfish." *Am. Fisher. Soc. Spec. Publ.* No. 5. Washington, D. C.
Cheng, T. C., Thakur, A. S., and Rifkin, E. (1969). Phagocytosis as an internal defense mechanism in the Mollusca: with an experimental study of the role of leucocytes in the removal of ink particles in *Littorina scabra* Linn. Proc. Symp. Mollusca, Part II, 546-563, *Marine Biol. Assoc.* India.

Cheng, T. C., Cali, A., and Foley, D. A. (1974). Cellular
    reaction in marine pelecypods as a factor influencing
    endosymbioses. In "Symbiosis in the Sea." (W. B. Vernberg,
    ed.). *Univ. South Carolina Press*, Columbia S.C. pp. 61-91.
Des Voigne, D. M. and Sparks, A. K. (1968). The process of wound
    healing in the Pacific oyster, *Crassostrea gigas*. *J.
    Invertebr. Pathol.*, 12, 53-65.
Feng, S. Y. (1959). Defense mechanism of the oyster. *Bull.
    N.J. Acad. Sci.*, 4, 17.
Feng, S. Y. (1965). Pinocytosis of proteins by oyster leucocytes.
    Biol. Bull., 128, 95-105.
Feng, S. Y. (1967). Responses of molluscs to foreign bodies, with
    special reference to the oyster. *Fed. Proc.*, 26, 1685-1692.
Haeckel, E. (1862). Die Radiolarien. Geo. Reimer, Berlin.
Pauley, G. B. and Heaton, L. H. (1969). Experimental wound repair
    in the fresh-water mussel *Anodonta oregonensis*. *J. Invertebr.
    Pathol.*, 13, 241-249.
Pauley, G. B. and Krassner, S. M. (1972). Cellular defense
    reactions to particulate materials in the California sea
    hare, *Aplysia californica*. *J. Invertebr. Pathol.*, 19,
    18-27.
Pauley, G. B. and Sparks, A. K. (1965). Preliminary observations
    on the acute inflammatory response in the Pacific oyster,
    Crassostrea gigas (Thurnberg). *J. Invertebr. Pathol.*,
    7, 248-256.
Polglase, W. J., Smith, E. L., and Tyler, R. H. (1952). Studies on
    human glycogen I. Preparation, purity, and average chain
    length. *J. Biol. Chem.*, 199, 97-104.
Reade, P. C. (1968). Phagocytosis in invertebrates. *Aust. J. Exp.
    Biol. Med. Sci.*, 46, 219-229.
Scott, R. M. (1969). Clinical analysis by TLC. *Ann Arbor-
    Humphrey Science Publ.* Ann Arbor, Michigan.
Sminia, T. (1972). Structure and function of blood and connective
    tissue cells of the fresh water pulmonate *Lymnaea stagnalis*
    studied by electron microscopy and enzyme histochemistry.
    Z. Zellforsch., 130, 497-526.
Stauber, L. A. (1950). The fate of India ink injected intra-
    cardially in the oyster, *Ostrea virginica* Gmelin. *Biol.
    Bull.*, 98, 227-251.
Staurt, A. E. (1968). The reticulo-endothelial apparatus of the
    lesser octopus, *Eledone cirrosa*. *J. Path. Bact.*, 96, 401-
    412.
Takatsuki, S. (1934). On the nature and function of the amoebocytes
    of *O. edulis*. *Quart. J. Microsc. Sci.*, 76, 379-431.
Tripp, M. R. (1958a). Disposal by the oyster of intracardially
    injected red blood cells of vertebrates. *Proc. Nat. Shell-
    fish. Assoc.*, 48, 143-147.
Tripp, M. R. (1958b). Studies on the defense mechanism of the
    oyster. *J. Parasitol.*, 44(Sect. 2), 35-36.

Wagge, L. E. (1951). The activity of amoebocytes and of alkaline
     phosphatases during the regeneration of shell of the snail
     Helix aspersa. *Quart. J. Microsc. Sci.*, 92, 307-321.
Wagge, L. E. (1955). Amoebocytes. *Int. Rev. Cytol.*, 4, 31-78.
Yonge, C. M. (1923). The mechanism of feeding, assimilation, and
     digestion in *Mya arenaria*. *Brit. J. Exp. Biol.*, 1, 15-63.
Yonge, C. M. (1926). Structure and physiology of the organs of
     feeding and digestion in *Ostrea edulis*. *J. Mar. Biol.*
     Assoc. U.K., 14, 295-388.
Yonge, C. M. and Nicholas, H. M. (1940). Structure and function of
     the gut and symbiosis with zooxanthellae in *Tridacna*
     crispata (Oerst). Bgk. Papers Tortugas Lab. Carnegie Inst.,
     32, 287-301.
Zacks, S. I. (1955). The cytochemistry of the amoebocytes and
     intestinal epithelium of *Venus mercenaria* (Lamelli
     branchiata), with remarks on a pigment resembling ceroid.
     Quart. J. Microsc. Sci., 96, 57-71.
Zacks, S. I. and Welsh, J. H. (1953). Cholinesterase and lipase in
     the amoebocytes, intestinal epithelium and heart muscle of
     the quahog, *Venus mercenaria*. *Biol. Bull.*, 105, 200-211.

Roles of <u>Mytilus</u> <u>coruscus</u> and <u>Crassostrea</u> <u>gigas</u> Blood Cells in Defense and Nutrition *

S. Y. Feng

J. S. Feng

MARINE RESEARCH LABORATORY AND
  BIOLOGICAL SCIENCES GROUP
UNIVERSITY OF CONNECTICUT
NOANK, CONNECTICUT

AND

T. YAMASU

TOMANO MARINE LABORATORY
OKAYAMA UNIVERSITY
SHIBUKAWA, TAMANO
OKAYAMA (706), JAPAN

* Contribution No. 111 from Marine Research Laboratory,
  University of Connecticut.

I. INTRODUCTION

Since the hay days of biology, hemolymph corpuscles of invertebrates have captured the attention of biologists. The voluminous
literatures published during the 1800s to early 1900s by men like
Cuénot, Metchnikoff, and others bear eloquent witness to their
pioneering effort in this field. This fascination of early
biologists with these cells has persisted to the present day.
In the past 15 years one has experienced an accelerated increase
in the output of reports dealing with invertebrate hemolymph
formed elements. A brief survey of the literature, however,
reveals that description of morphological types remains to be
the major theme of most published reports. This may have been the
fact that the diversity of forms of invertebrate hemolymph formed
elements merely reflects the richness of invertebrate species.
Thus, the task of describing and cataloging of invertebrate
hemolymph formed elements must be continued, because it is a
prerequisite to the development of any unifying concepts on the
evolution and function of these cells. In our opinion, there are
still major gaps in our knowledge regarding these cells to warrant
an meaningful comprehensive review at this time. However, some
generalizations may be made regarding the lamellibranch amoebocytes
which perhaps represent the best known case.

Morphologically the lamellibranch amoebocytes may be divided
into two basic types: (1) agranulocytes containing a few, if any,
cytoplasmic granules, and (2) granulocytes showing numerous
granules of different sizes and staining characteristics in the
cytoplasm. Such a scheme has been used to classify various
lamellibranch amoebocytes: *Ostrea edulis* (Takatsuki, 1934),
*Anomia liscki, Arca inflata, Glycymeris vestitus, Libitina
japonica, Mytilus edulis* and *Ostrea circumpicta* (Ohuye, 1937,
1938a,b), *Crassostrea gigas* (Ruddell, 1971a,b), *Crassostrea
virginica* (Feng, *et al.*,1971; Foley and Cheng, 1972), and
Placopecten magellanicus (Stevenson and South, 1975).

The multiple functions of lamellibranch amoebocytes have
been well documented in the literature. The reviews of Takatsuki
(1934), Younge (1946), Wagge (1955), Owen (1966), and Purchon
(1968) stress the importance of these cells in digestion, while
those of Stauber (1961), Tripp (1963; 1970), Cheng (1967), and
Feng (1967) emphasize the role of amoebocytes in defense. It is
clear that the underlying mechanism of digestion and defense is
associated with the ability of amoebocytes to recognize,
phagocytose, and digest foreign particles whether be it food,
e.g., algal cells, or invading microorganisms, e.g., bacteria.
The two processes differ only in the final fate of the ingested
particles. Yonge (1926) suggests that food particle-laden
amoebocytes in the lumen of the intestine migrate back across
the intestinal epithelium into the deep tissues where the
ingested food particles are digested intracellularly; this view

has been challenged by George (1952). It is significant to note
that the attrition of amoebocytes laden with bacteria takes
precisely the reverse course of the above route (Tripp, 1960).
In their review of the mechanism of excretion, Wagge (1955) and
Potts (1967) again have noted that particle-laden amoebocytes
of the circulatory system are eliminated by the route of the
intestine, pericardial cavity, nephredia, or mantle cavity and
return after releasing the particles. Based upon these data,
Potts (1967) boldly considers the amoebocytes as being analogous
to the reticulo-endothelial system of vertebrates. It has also
been demonstrated that the lamellibranch amoebocytes possess a
sundry of intracellular enzymes which hydrolyze a variety of
substrates of carbohydrates, proteins, and lipids (Yonge, 1926,
1937, 1946; Nelson, 1933; Takatsuki, 1934; George, 1952; Zacks
and Welsh, 1953; Zacks, 1955; Wagge, 1955; Janoff and Hawrylko,
1963; Eble, 1966; Feng *et al.*, 1971; Narain, 1972; Cheng and
Rodrick, 1974; Cheng *et al.*, 1975). These enzymes are known to
be of lysosomal origin in vertebrates in which the role of defense
against microbial invaders is assigned exclusively to cells of
the reticulo-endothelial system. Lamellibranch amoebocytes are
known to participate in digestion and defense. In actuality,
one finds it extreme difficult to draw a well-defined line
between the two functions. It may be speculated that originally
phagocytosis was a means of gathering, processing, and assimilating
of food particles and that its role in defense was developed
secondarily during evolution.

In addition to their role in digestion, defense, and excretion
mentioned above, lamellibranch amoebocytes also play important
roles in storage of waste products, e.g., ceroid bodies (Zacks,
1955), shell formation (Wagge, 1955; Wilbur, 1964), and wound
repair (Ruddell, 1971a,c).

In this report we will present a comparative morphological
study on the amoebocytes of *Crassostrea gigas* and *Mytilus
coruscus* with emphasis on their roles in defense and nutrition.

II. MATERIALS AND METHODS

Oysters and Mussels. Japanese oysters, *Crassotrea gigas*,
and mussels, *Mytilus coruscus*, were collected from Ushimado and
Wasa Island, Seto Inland Sea, Okayama, Japan, respectively, during
February and March, 1973. The animals were maintained in a running
seawater trough (13°C) during 1½ months of studying period.
Hemolymph was taken from the adductor muscle using a syringe of
2.5 or 10 ml capacity fitted with a 26-gauge hyperdermic needle
(Feng *et al.*, 1971).

Light microscopy. Oyster blood smears were fixed briefly in
absolute methanol and stained for 12 min with Giemsa stain
(1 part in 9 parts of 0.4 M phosphate buffer, pH 6.4). It was
discovered in a preliminary study that *M. coruscus* amoebocytes

were extremely sensitive to methanol fixation. Mussel blood smears,
therefore, were fixed in 10% formalin for 15 min followed by Giemsa
staining.

Electron microscopy. Pellets of amoebocytes were prefixed in
1.5% glutaldehyde, pH 7.4, at room temperature, 10-15°C, for 4 to
6 hr followed by postfixation in 1% OsO$_4$ for 2 hr in an ice bath
at 0.5°C. The fixed specimens were then washed three times in
0.1 M cacodylate buffer, pH 7.4, with 8.5% sucrose, dehydrated at
0°C, and embedded in Epon.

Sections cut on a Sorvall MTB 2 Ultratome with a glass knife
were contrasted with uranyl acetate and lead citrate; they were
mounted on the grids with formvar film and examined in a Hitachi
HS8 electron microscope operating at 50 kV.

Phagocytosis *in vitro*. *In vitro* phagocytosis was performed
at room temperature (10°C-15°C). Five ml of hemolymph was drawn
from one oyster and one mussel; each was mixed gently with 1 ml
of *Escherichia coli* suspension in sea water in a large test tube.
After a 90-min incubation, two samples were taken. One was processed for electron microscopy in the manner described above; the
other portion was processed for demonstration of acid phosphatase
in these cells according to the procedure of Farquhar *et al.*
(1972) using a modified Gomori medium, pH 5.0, with β-sodium
glycero-phosphate as substrate. The controls were incubated
without substrate.

Extraction of pigments. Pooled hemolymph was centrifuged at
3000 rpm for 10 min and the supernatant was decanted. The yellow-
ish cell pellet obtained was repeatedly extracted for pigments
with 5% HCl in methanol or acetone; each time cells were resuspended in the extractant and centrifuged until the pellet rendered
free of all pigments. The supernatant was then diluted with equal
volume of water and the pigments were transferred to petroleum
ether. After repeating this procedure three times, the pigments
in petroleum ether was washed with 500 ml of water and dried over
anhydrous sodium sulfate. The dried solution was concentrated
rapidly in a rotary evaporator with a cold trap to a few ml *in
vacuo*. The concentrated pigment extract was redissolved in a
few ml of petroleum ether, transferred to a screw cap test tube
dried in a stream of nitrogen, and kept frozen (-20°C) until such
time when it is needed.

Thin layer chromatography. Separation and identification of
the pigments were carried out by thin layer chromatography. The
nitrogen dried pigment extract was dissolved in 10% acetone in
petroleum ether, spotted on thin layer plates of silica gel F-254
(Merck, layer thickness, 0.25 mm) and developed with a mixture of
reference standards (retinol, β-carotene, echinenone, canthaxanthin,
zeaxanthin, cholesterol and palmitic acid in 10% acetone in
petroleum ether) for 20 min at 22°C in a solvent of 10% acetone in
petroleum ether. The developed chromatogram was examined for
colored spots by visual inspection under ultraviolet light (360 nm)

and sprayed with the Carr-Price reagent (a saturated solution of
$SbCl_3$ in chloroform).

III. OBSERVATIONS

A. LIGHT MICROSCOPY

   Crassostrea gigas amoebocytes. Oyster amoebocytes on Giemsa-
stained blood smears can be differentiated into three types based
on their nuclear morphology; each type comprises both agranular
and granular cells. Type I amoebocytes are characterized by a
small (3-5 µm) basophilic compact nucleus surrounded by a moderate
amount of cytoplasm (Fig. 1a,b). In Type II amoebocytes the
nucleus is larger (5-8 µm), but less compact; it is round to oval
in shape and stained pinkish purple (neutral) (Fig. 1c,d). Type
III amoebocytes have a large (6-10 µm) esosinophilic nucleus of
fine outline and a large amount of cytoplasm (Fig. 1e,f).
   Mytilus coruscus amoebocytes. Based on nuclear morphology,
mussel amoebocytes can be grouped into four types; each type
comprises agranular and granular cells. The nucleus of Type I
(Fig. 1g,h) is basophilic, round, and the smallest (3-4 µm)
of all types. Cells are generally round to square in shape with
moderate amount of cytoplasm and show little variation in size
(9-11 µm). They exhibit minimum dispersion on the slide and
remain largely aggregated. In Type II (Fig. 1i,j) the nucleus is
slightly larger (4-5 µm) and characterized by its compact dense
basophilic appearance. The less dense basophilic round to oval
shaped nucleus of Type III (Fig. 1k,l) has a range of 5-8 µm in
size. Type IV is characterized by an eosinophilic oval nucleus
(5-10 µm) (Fig. 1m,n).
   The cytoplasm of mussel amoebocytes is slightly eosinophilic.
In the granulocytes (Fig. 1h,j,l,n) varying numbers of refractile
or/and nonrefractile eosinophilic granules, coarse or fine, are
seen. Granulocytes constitute an average of 66.6% (SD=11.3,
SE=3.4 n=11) of the total cell population.
   The basic difference in the cytoplasmic granules between the
two bivalve molluscs is their apparent susceptibility or resistance to methanol. While the cytoplasmic granules of *C.
gigas* leucocytes are preserved by the absolute methanol, the
granules of *M. coruscus* are extremely sensitive to this fixative.
Such difference presumably reflects the chemical nature of the
granules. It is also noted that the mussel amoebocytes are much
less amoeboidal than the oyster amoebocytes.

B. ELECTRON MICROSCOPY

   Crassostrea gigas amoebocytes. With the exception of granules,
the organelles in agranular (Fig. 2) and granular (Fig. 5) cells
of *Crassostrea gigas* are similar. Round- to oval-shaped nuclei

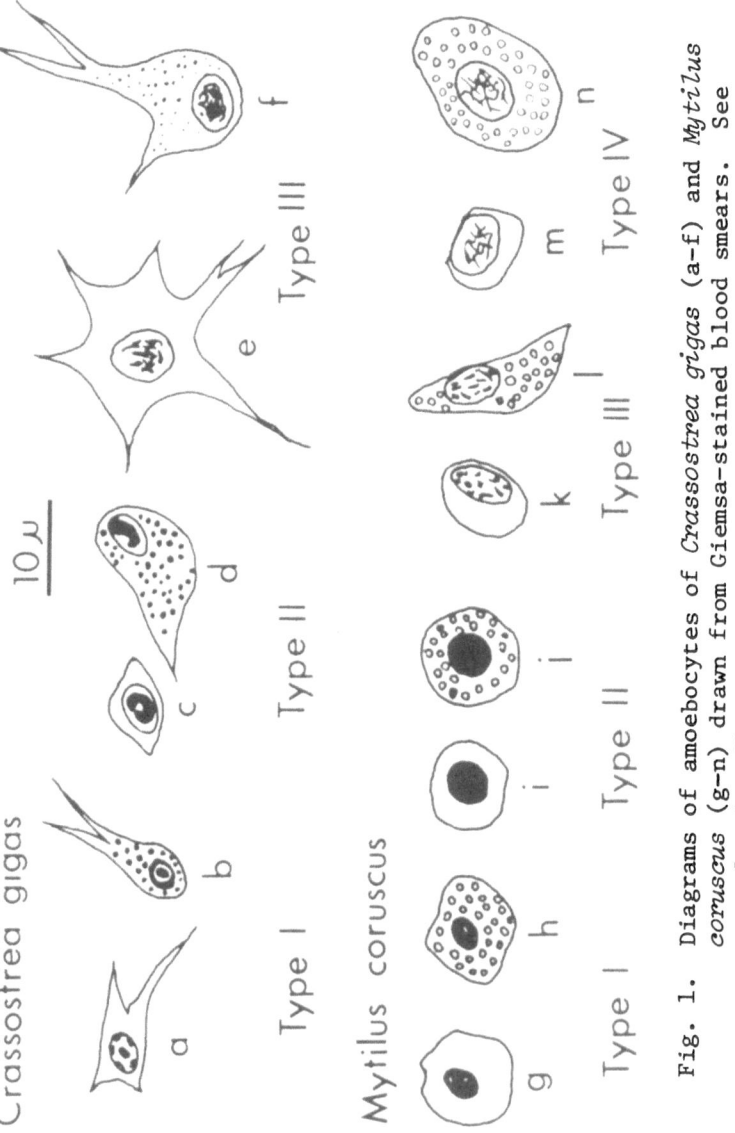

Fig. 1. Diagrams of amoebocytes of *Crassostrea gigas* (a-f) and *Mytilus coruscus* (g-n) drawn from Giemsa-stained blood smears. See text for details.

have chromatin which forms a thin layer along the inner nuclear
membrane and clumps within. A nucleolus may be seen in some cells.
Occasionally there are tiny vesicular protrusions between the
inner and outer nuclear membrane. The outer nuclear membrane
often bear ribosomes. Nuclear pores, however, are rarely observed.
In accordance with the observation of light microscopy binucleated
cells are encountered (Fig. 3).

The oval- or rod-shaped profiles of mitochondria, which normally measure less than 1.7 μm, exhibit localized or random distribution. Tubular cristae, a fairly dense matrix, and fine
particles resembling ribosomes, which are located in the matrix
and on the cristae, can be identified in the organelle. Both
smooth and rough endoplasmic reticulum are present in varying
amounts in the amoebocytes. While short cisternae or networks
of such cisternae may be found in an agranular cell or in a cell
with very few granules, they are largely vesicular in granular
leucocytes. Free ribosomes and polyribosomes are also present
but sparsely in most cells.

Golgi apparatus, though small, ca. 2 μm, is found quite
frequently. Centrioles with small numbers of microtubules are
often located near Golgi apparatus or nucleus. In few cells
parallel arrays of microfibrils are also present near a nucleus
(Fig. 4).

In granular amoebocytes, several types of granules may be
recognized. Normally a granular leucocyte contain only one type
of granules although some have two types. Size variations of the
granules are summarized in Table 1.

Type A granules (Fig. 2,3) are irregularly shaped and have
a wide range of 0.32-1.91 μm in size (mean, 0.86 μm; SD, 0.268;
n=357). Generally they are electron-light and contain sparse
or moderate amount of amorphous materials within a limiting
membrane. In some, however, the intragranular contents also
include tiny vesicles, dense particles, or/and a complete or incomplete thin layer of moderately dense material closely associated
with the membrane. Of all the granules this type is most often
encountered. Basophilic granules described by Ruddell (1971b)
appear to resemble this type.

The ellipsoidal Type B granules (Fig. 5) range from 0.38-
0.90 μm in size (mean, 0.60 μm; SD, 0.122; n=31). They possess
a narrow dense band which lines closely to the membrane and
encloses a lighter interior containing amorphous material.

Type C' granules (Fig. 6) are more or less spherical
(0.14-0.67 μm; mean, 0.34 μm; SD, 0.126; n=100), and characterized
by their very dense content which fills the whole granule and
tends to obscure the wavy limiting membrane. Granules of
acidophiles reported by Ruddell (1971b) probably correspond to
this type.

Within the same cell where Type C' granules are found there
are granules which also contain a dense core but surrounded by an

Fig. 2.   An oyster agranular amoebocyte. Conventional organelles
          such as nucleus with its double membrane and chromatin,
          mitochondria, a small amount of short cisternae of
          rough endoplasmic reticulum as well as free ribosomes
          can be identified. X 13,200.

Fig. 3.   A binucleated oyster granular amoebocyte containing
          Type A granules. X 8,300.

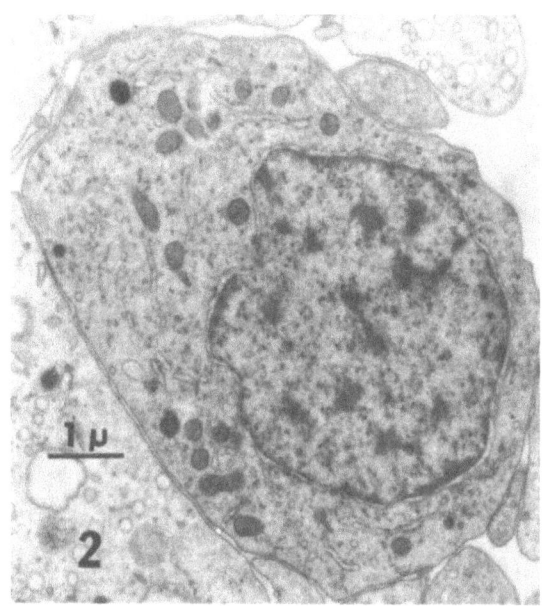

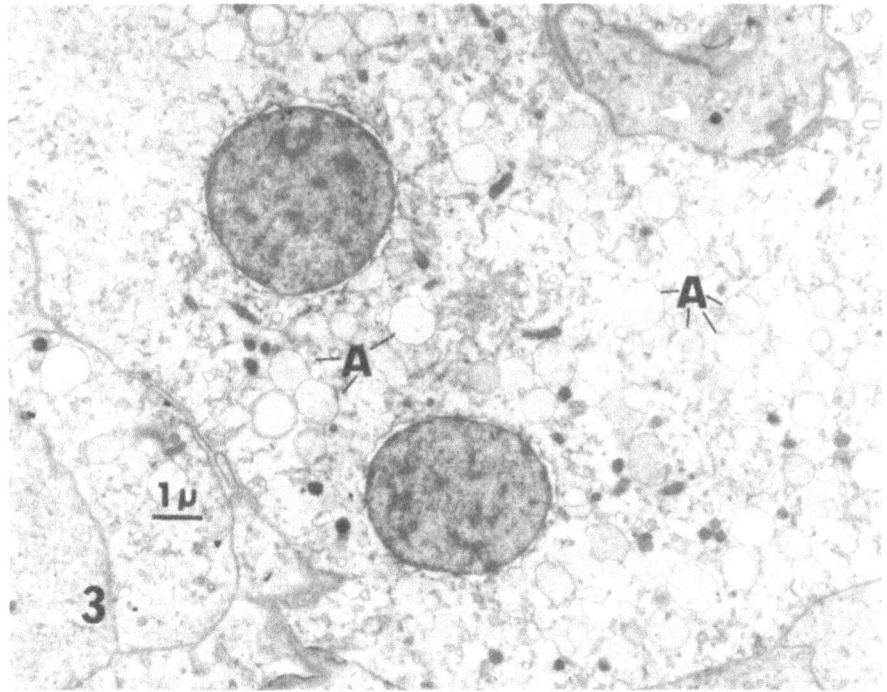

Fig. 4. An oyster granular amoebocyte with a small quantity of
         intracytoplasmic fibrils (Mf) near the nucleus. Type
         A granules (A), Golgi apparatus (G), and rough endo-
         plasmic reticulum are also seen. X 18,000.

Fig. 5. An oyster granular amoebocyte showing Type B granules
         (B) and other organelles. Both smooth and rough endo-
         plasmic reticulum are in vesicular form. X 21,000.

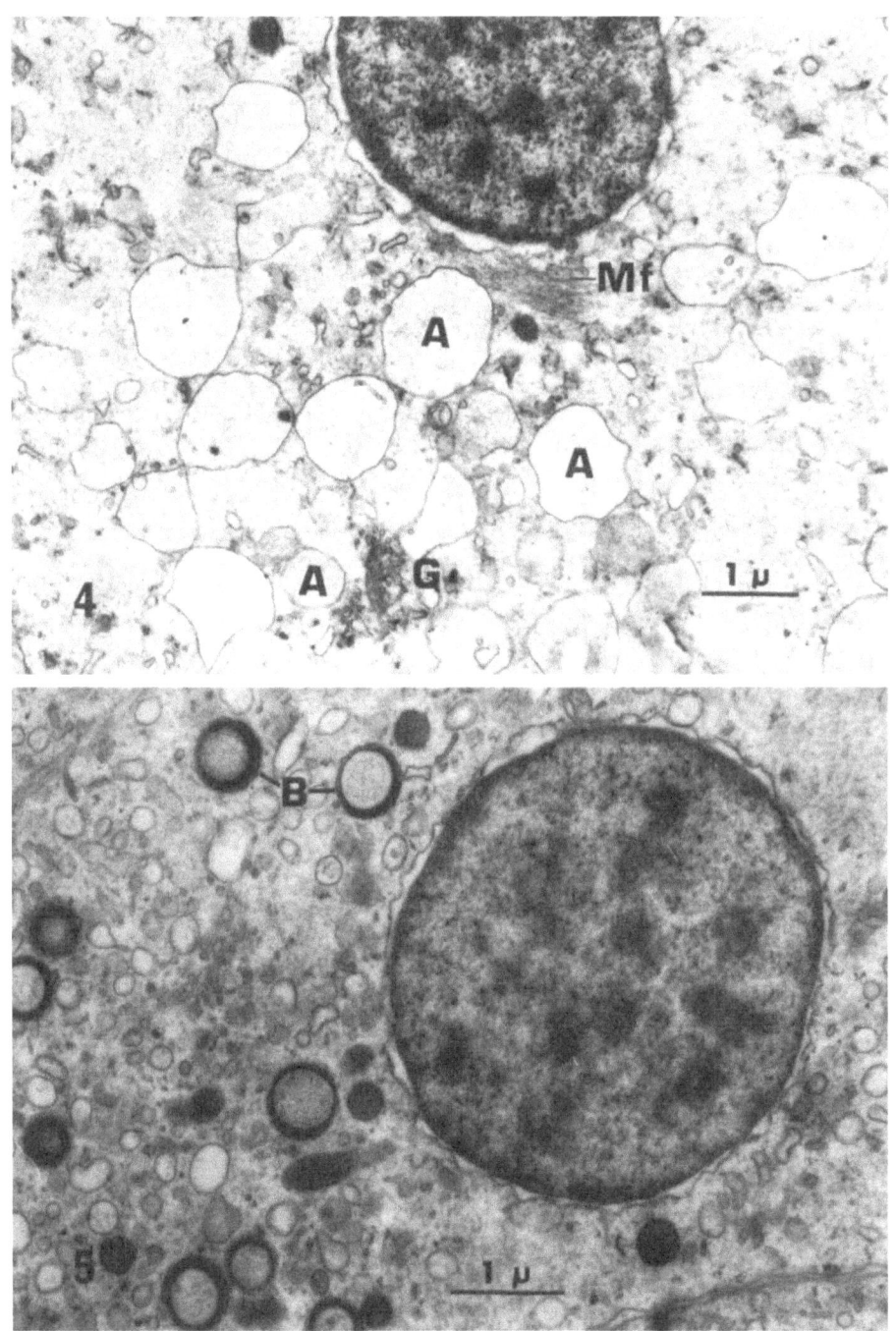

electron lucent or a moderately dense area may also be present
(Fig. 6). Although these granules appear to be the developmental
stages of Type C' granules further study is needed to varify this
view.

An oval-shaped organelle (0.7-2.0 μm), which is filled with
moderately osmiophilic, finely granulated material and may appear
homogeneous, is probably lipid bodies (Fig. 8).

Following an exposure of the amoebocytes to *E. coli*, phagocytic vacuoles with ingested bacteria are found in both agranular
and granular cells (Fig. 7,8) which show bacteria bearing phagosome with cytoplasm granules either in juxtaposition or containing
materials from cytoplasmic granules. Acid phosphatase reaction
product resembling core materials of Type C' granules and occasional
lipid like bodies are present in phagosomes containing ingested
bacteria (Fig. 8). It appears that the cisternae and associated
vesicles of the Golgi apparatus also contain acid phosphatase
reaction products.

Mytilus coruscus amoebocytes. In both agranular (Fig. 9) and
granular (Figs. 10,11, 12) amoebocytes, round- or oval-shaped
nuclei are seen with their double membrane, chromatin clumps, and,
sometimes, nucleoli and nucleopores. Microtubular mitochondria
(Fig. 10) show random or localized distribution. While vesicles
and short narrow cisternae of endoplasmic reticulum are present,
they are generally inconspicuous in granular amoebocytes; sparse
or moderate amount of relatively long larger cisternae of rough
endoplasmic reticulum are found in agranular amoebocytes. Small
Golgi apparatus and centriole, frequently located in close proxi-
mity (Fig. 11), are seen fairly often. In addition, vacuoles with
mitochondria, cytoplasmic granules, vesicles of endoplasmic reticulum, fine particles, and amorphous material are also present in some
cells suggesting the presence of autophagic activities within
these cells.

Several prominent types of granules are observed in the granular amoebocytes. They are all limited by a membrane which encloses various intragranular contents. As shown in Figure 10,
amoebocytes with mixed granules are also of common occurrence.
Variations in the size of these granules are shown in Table 2.

The intragranular content of Type A granules (Fig. 10)
(0.45-1.1 μm) is largely fine particles of light density and may
appear somewhat homogeneous. They are the lightest of all
granular types.

Type B granules (Fig. 10) (0.45-1.64 μm) are spherical and
possess a very dense round core embedded in a fairly dense matrix
of finely granulated material. At times their wavy outer membrane appear double-layered (Fig. 10). Type B granules occur
more frequently than other types.

Type C granules (Fig. 10) are typically irregular in their
outlines and range widely in their size (0.72-2.29 μm). They
also possess a very dense core which is centrally or accentrically

Table 1. The size variation in the three types of *Crassostrea gigas* amoebocyte granules revealed in electron micrographs.

Type of granules	$\overline{X}$	s^2	SD	SE	n
A	0.86	0.0718	0.268	0.014	357
B	0.60	0.0148	0.122	0.022	31
C'	0.34	0.0158	0.126	0.013	100

$\overline{X}$ is expressed as the mean diameter of granules in µm.

Table 2. The size variation of *Mytilus coruscus* amoebocyte granules revealed in electron micrographs.

Type of granules	$\overline{X}$	s^2	SD	SE	n
A	0.89	0.0471	0.217	0.033	43
B	0.85	0.0846	0.291	0.036	66
C	1.34	0.1536	0.392	0.080	24
D	1.90	0.2016	0.449	0.159	8
E	0.35	0.0034	0.059	0.001	65
L	0.66	0.0789	0.281	0.029	94

$\overline{X}$ is expressed as the mean diameter of granules in µm. L denotes lipid-like bodies.

Fig. 6. An oyster granular amoebocyte containing Type C'
        granules (C'). Note also the presence of centriole
        (Ce), Golgi apparatus (G),and relatively large mito-
        chondria. X 24,500.

Fig. 7. This electron micrograph illustrates phagocytosis of
        bacteria by an oyster granular amoebocyte under natural
        condition, i.e., bacteria are not introduced experi-
        mentally into the oyster. Phagocytic vacuole (V),
        Type B granules (B), Golgi apparatus (G), vesicles of
        rough and smooth endoplasmic reticulum are seen.
        X 20,500.

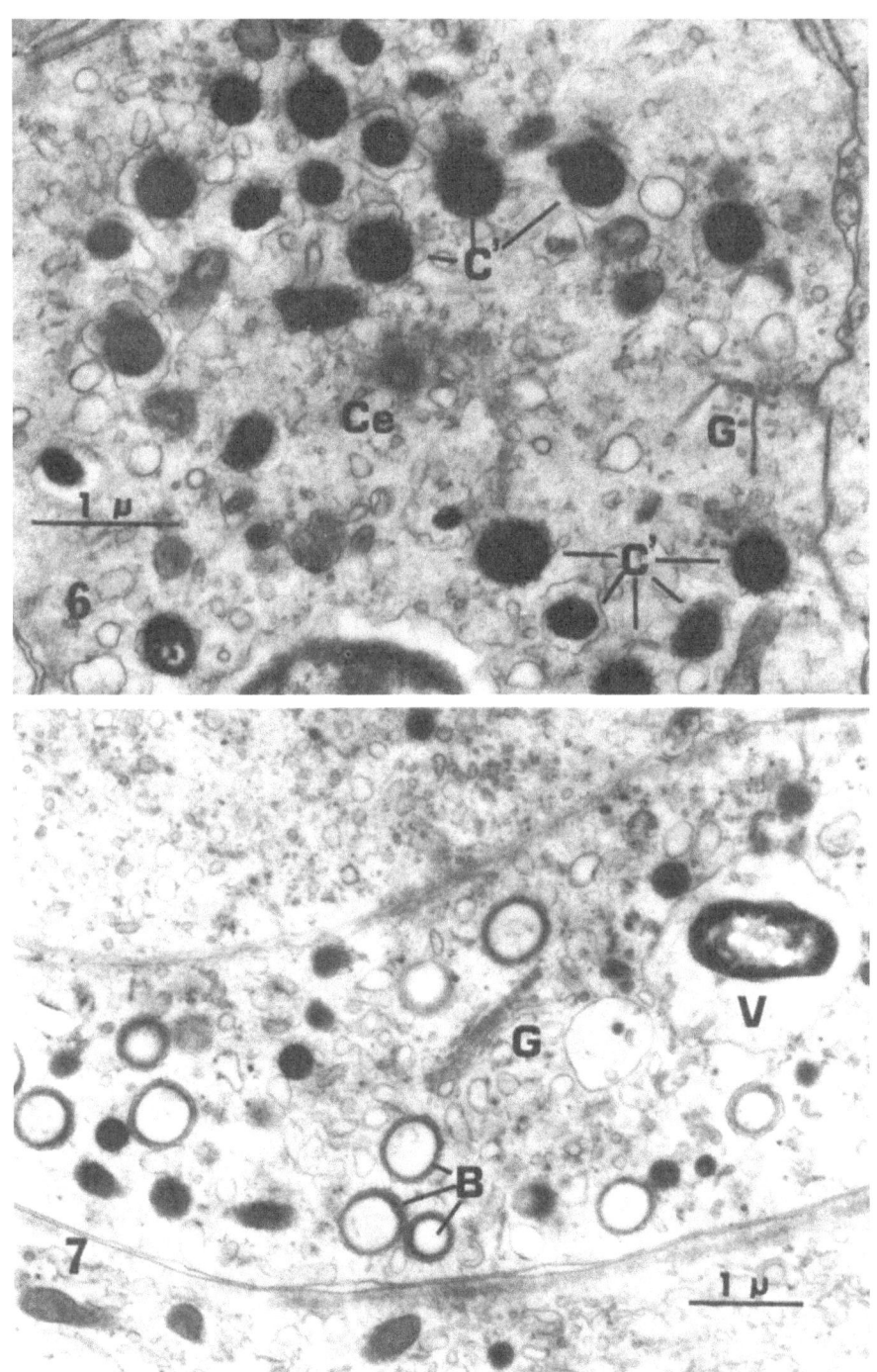

Fig. 8. Acid phosphatose preparation of an oyster granular
         amoebocyte demonstrating several phagocytic vacuoles
         with ingested bacteria that were introduced experi-
         mentally using an *in vitro* system described in the text.
         One such vacuole (V-a) also has a large lipidlike body
         (L), while two others (V-b and V-c) contain lead
         deposit (arrows) indicating the presence of acid
         phosphatose activities. Lead deposites are also seen
         to be associated with Golgi cisternae (Gc). The
         bacteria in vacuole labelled as V-d, shows distortion
         which may suggest degradation. Specimens were pro-
         cessed according to the method of Farquhar *et al.*,
         (1972). Sections stained in uranyl alone. X 20,400.

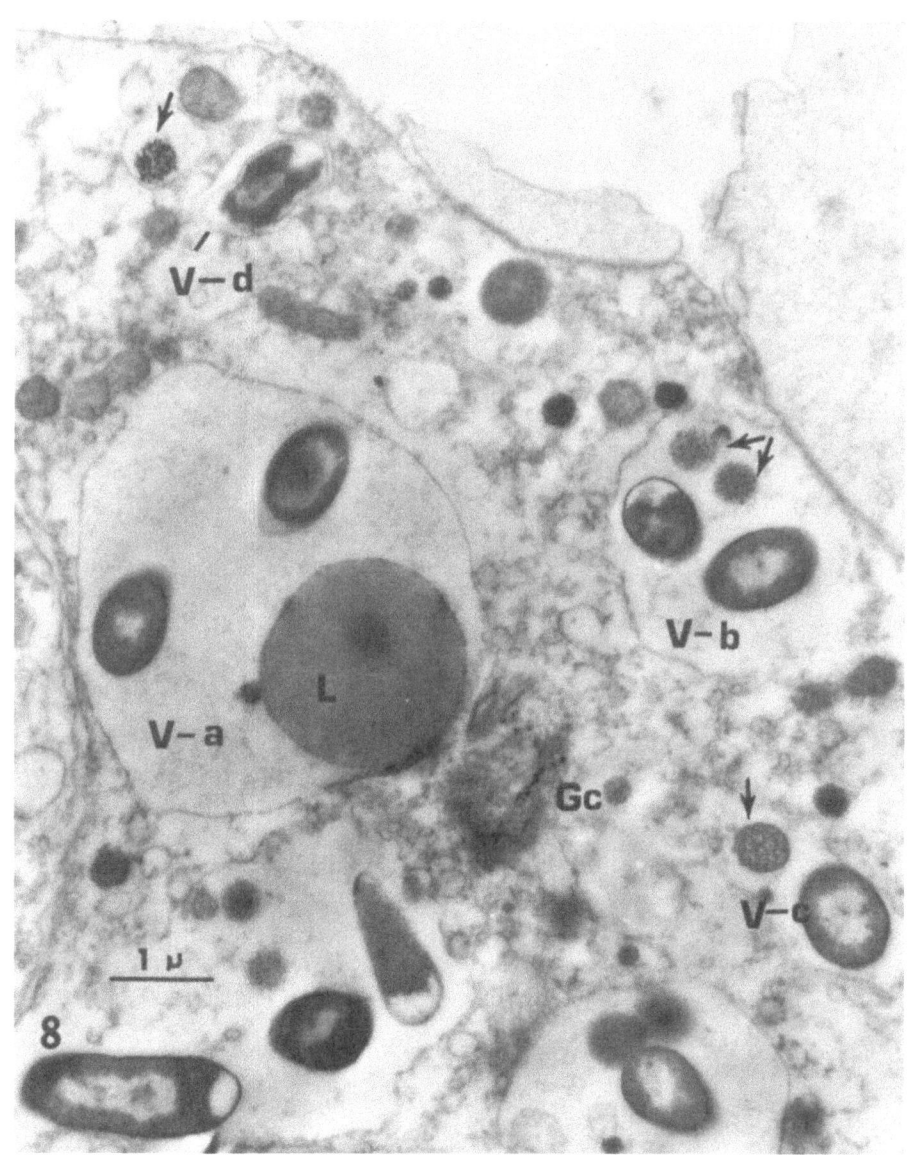

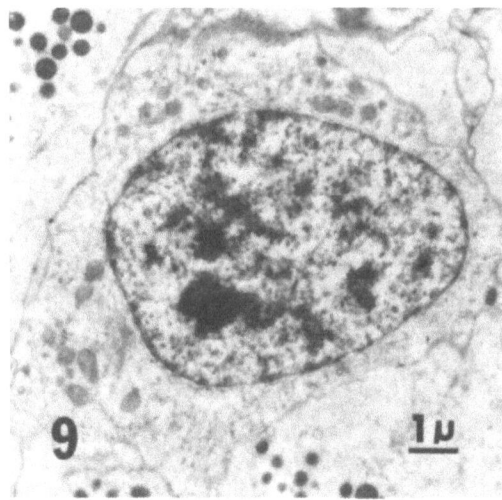

Fig. 9. A mussel agranular amoebocyte. Note a relatively large
         nucleus with its double membrane and chromatin clumps.
         Other organelles seen are mitochondria, short cisternae
         of rough endoplasmic reticulum, and free ribosomes.
         X 7,500.

located. The area between the core and the outer membrane, however,
is largely translucent with or without sparse to moderate amount of
coarse granulated material.

Type D granules (Fig. 10) are characterized by their large
size (1.71-2.45 µm) and the more or less ellipsoidal shape of
both their outline and dense core. The area between the membrane
and the core is more or less filled with coarse particles.

Type E granules, the smallest in size (0.18-0.47 µm), are
spherical dense bodies (Fig. 11). If present, they generally
constitute the major type within an amoebocyte. Lipidlike bodies
are dense organelles ranging from 0.27-1.63 µm in size and can be
found in large numbers in a cell (Fig. 12). They are generally
spherical although few exhibit irregular shape.

Phagocytosis of *E. coli* was observed in both agranular and
granular amoebocytes (Fig. 13). Ingested bacteria may or may
not be enclosed within a vacuole. In either case dense acid
phosphatase reaction products are often seen surrounding the
bacterial cell wall; also fusion of Type C granules with phagocytic
vacuoles is also evident in Figure 13. In 90-min sample, cell wall
of most of the ingested bacteria showed little structural change,
while cytoplasm exhibited some disintegration. Occasionally
vacuoles which appear to contain disintegrated bacteria were seen.
Infrequently, a phagocytic vacuole is found to be within another
vacuole (Fig. 13). In addition to bacteria, phagocytic vacuoles
may also contain vesicles of endoplasmic reticulum or/and dense
spherical bodies as well as granulated materials very much like
the cores and matrix of Type C and D granules.

C. THIN LAYER CHROMATOGRAPHY

The results summarized in Figure 14 clearly suggest the presence of carotenoids and flavonoids in the extract of amoebocytes
from both species. The characteristic blue and green fluorescence,
and dull brown color spots revealed by exposing the plate to uv
irradiation probably signify the presence of 3, 5 methoxylated
flavonols, flavonal without a 5-OH group and flavonol 3-glycosides,
and flavones, respectively, in the amoebocyte extract (Fig. 14A).
Interspecific qualitative differences in flavonoids are also
apparent; the extract from oyster amoebocytes seems to contain a
preponderance of 3, 5-methoxylated flavonols, coumarins, and
cinnamic acids, while the mussel cell extract contains approximately equal amounts of flavonol without a 5-OH group and 3,5-
methoxylated flavonols. Since no known standards of flavonoids
was cochromatographed with the sample, the results presented here
which are based largely upon the characteristic fluorescence, must
be regarded as preliminary.

In both *C. gigas* and *M. coruscus* samples, five well separated
spots are noted when the chromatogram was sprayed with Carr-Price
reagent. The fastest moving spots with a Rf value of 0.58-0.59

Fig. 10. A mussel granular amoebocyte with various types of
          granules which are labelled as A, B, C, and D. A
          centriole (Ce) is also present. X 13,700.

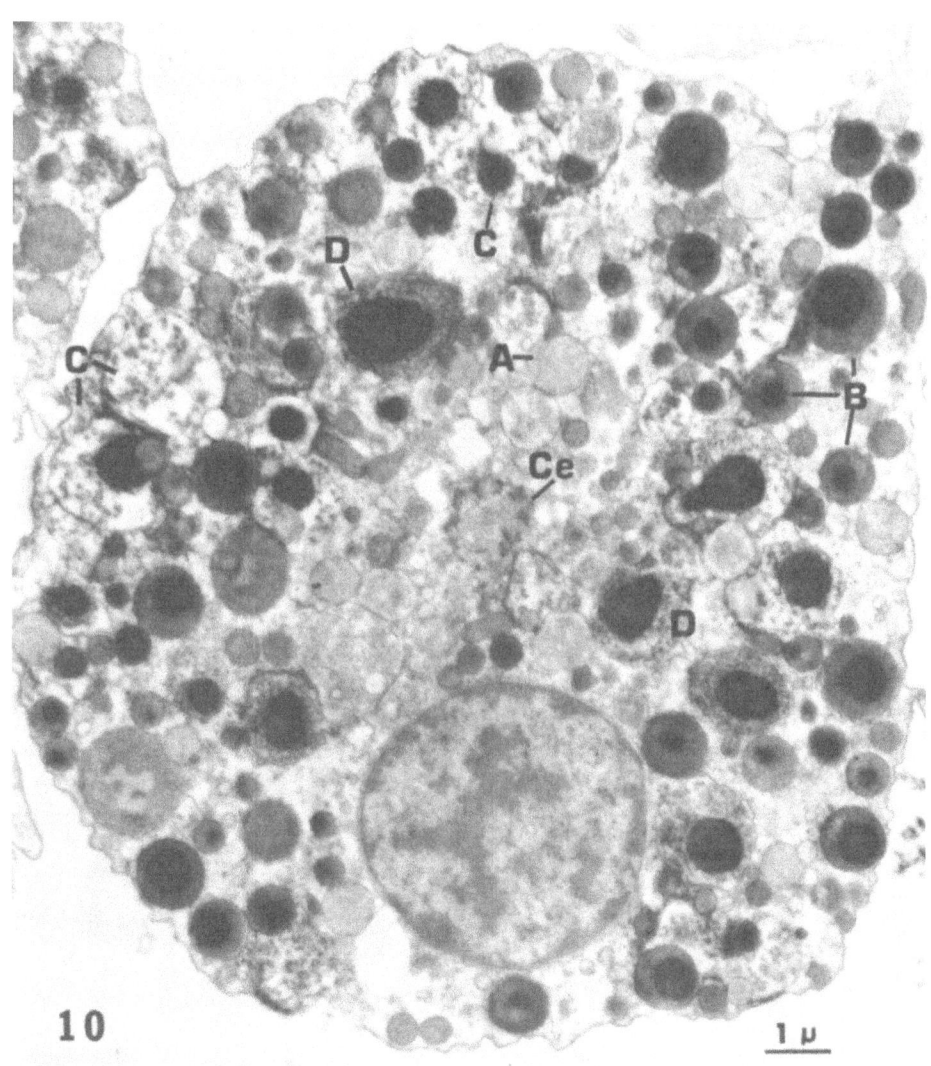

Fig. 11. A mussel granular amoebocyte displaying Type E granules
          (E), centriole (Ce), Golgi apparatus (G), mitochondria,
          vesicular form of rough endoplasmic reticulum. X 20,000.

Fig. 12. A mussel granular amoebocyte shown here is filled with
          lipid-like bodies (L). Also note the presence of
          closely situated Golgi apparatus (G) and centriole
          (Ce). Vesicles of rough endoplasmic reticulum are
          relatively fewer than other cell types.

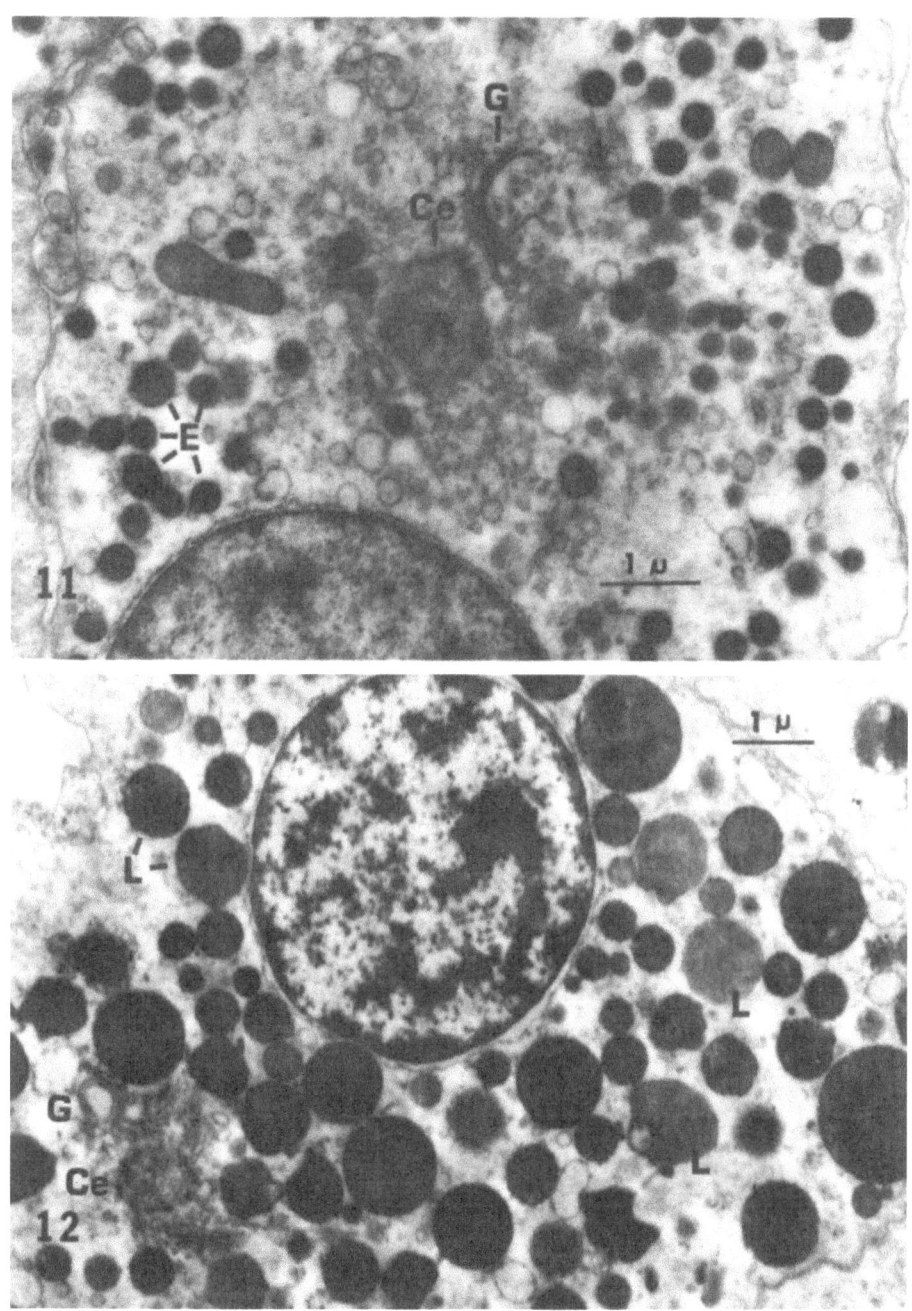

Fig. 13.   Acid phosphatase preparation of a mussel granular
           amoebocyte containing mainly Type B granules (B).
           It shows two phagocytic vacuoles with ingested bacteria
           (V-a and V-b). V-a appears to be within another
           vacuole (long arrow) and exhibit acid phosphatase
           reaction products (arrow heads) on the vacuolar
           membrane as well as the cell wall of bacteria. Note
           a large vacuole (Lv) which appears to have undergone
           fusion with several Type C granules (C), and is
           apparently being fused with two more Type C granules
           (short arrows). Heavy acid phosphatase reaction
           products are seen to coat the cell wall of bacteria.
           Specimens were processed according to the method of
           Farquhar *et al.*, (1972). Sections stained in uranyl
           alone. X 19,000.

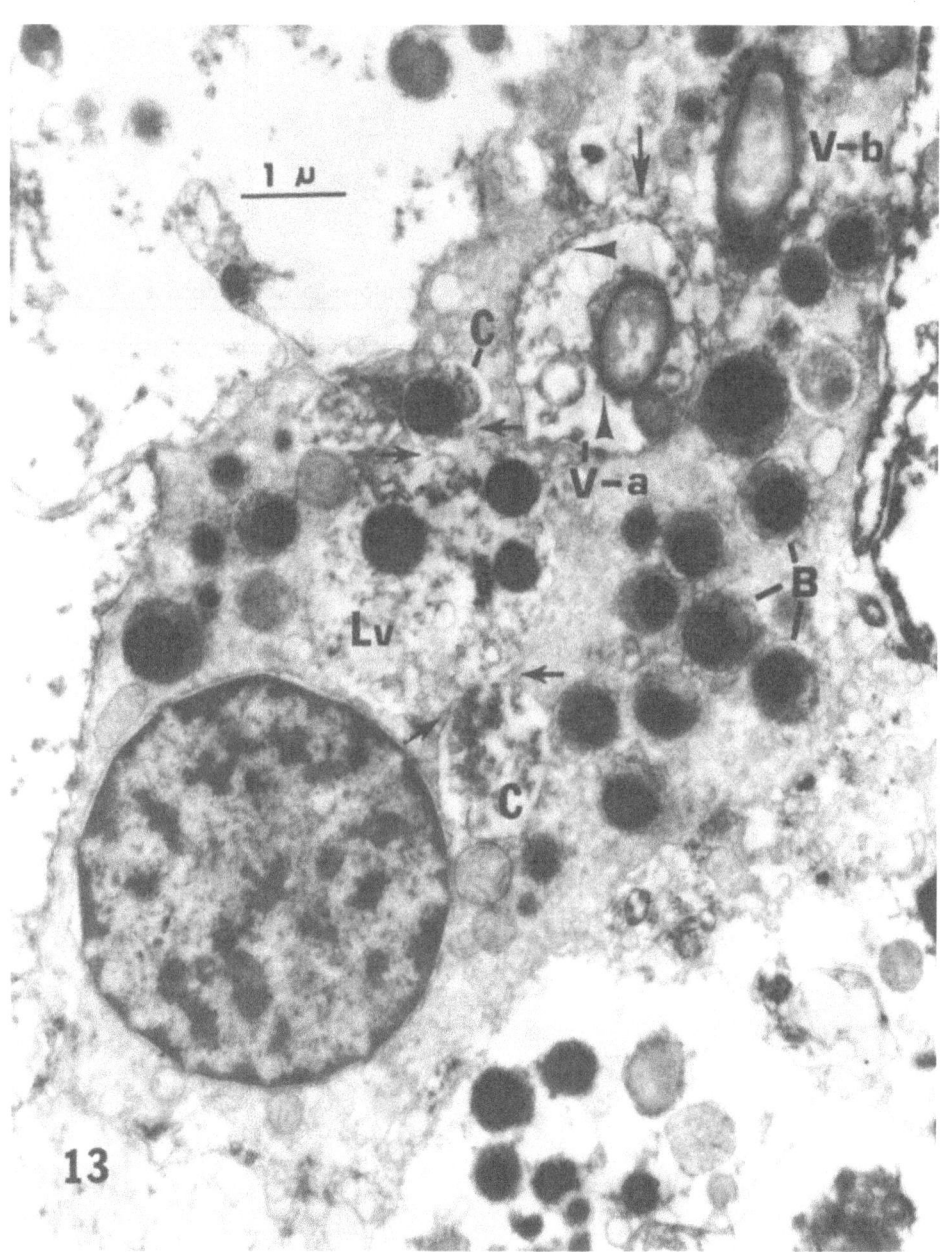

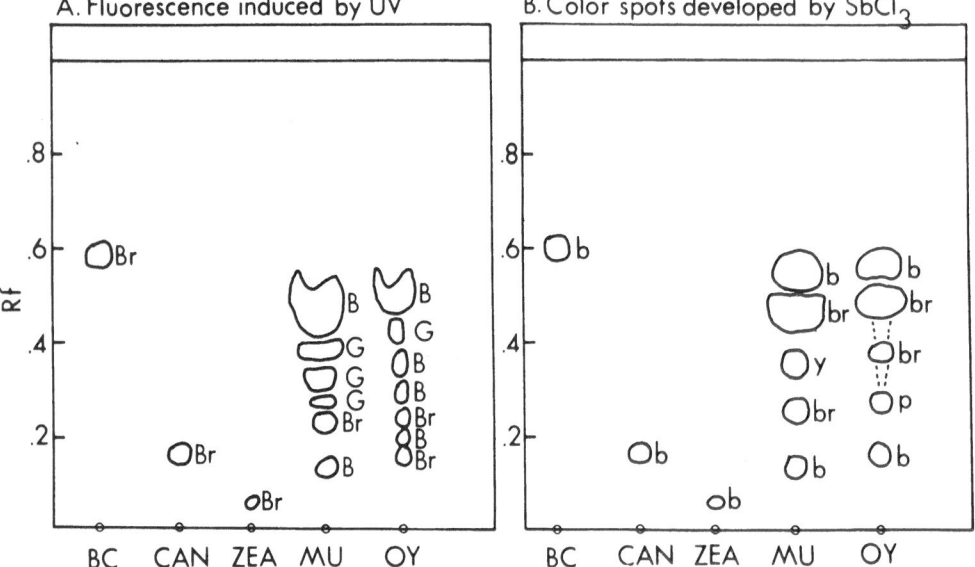

Fig. 14. Thin layer chromatogram of carotenoids and flavonoids
          from extracts of *Crassostrea gigas* (OY) and *Mytilus
          coruscus* (MU) amoebocytes. Reference standards are
          β-carotene (BC), canthaxanthin (CAN), and zeaxanthin
          (ZEA). Br, B, and G denote brown, blue, and green
          fluorescence, respectively, when the plate was exposed
          to long wave length uv light (A). Blue (b), brown
          (br), yellow (y), and pink (p) indicate characteristic
          color spots after the chromatogram was treated with
          Carr-Price reagent (B).

approximate that of the β-carotene standard (0.60), while the
slowest moving spots show Rf values of 0.15 and 0.17, respectively,
which correspond to that of canthaxanthin standard (0.17) (Fig.
14B). The three remaining spots having intermediate Rf values
between β-carotene and canthaxanthin are not identified owing to
the lack of proper standards. It is interesting to note that
this region also exhibits interspecific variation in terms of
differences in color reactions to the spray reagent in certain
spots.

IV. DISCUSSION

In the study of the ultrastructure of *C. gigas* amoebocytes,
we recognize two major cell types: agranular and granular cells,
which generally agree with the observations of Ruddell (1971a,b).
He recognizes two types of cytoplasmic granules in the granulocytes: acidophilic and basophilic granules. Based upon their size
and characteristic morphological features, these granules correspond Type C' and A granules, respectively, reported in the present
study. A third granule designated as Type B has also been described for *C. gigas* amoebocytes. A comparison of the known
granular types from *C. gigas* and *C. virginica* with respect to
their sizes, is summarized in Table 3. Apparently granular
amoebocytes of *C. gigas* and *C. virginica* contain both Type A and
B granules; the morphology and size of these granules are quite
similar (P>0.10∿0.50). Type C granules occur only in *C. virginica*
amoebocytes, while Type C' granules are associated with the amoebocytes of *C. gigas*.

It has been suggested that during encapsulation (Rifkin *et al.*,
1969) and wound repair (Ruddell, 1971a), undifferentiated agranular
amoebocytes may serve as the precursor of the fibroblastlike
cells and myoblasts which are characterized by the presence of
intracytoplasmic fibrils and muscle filaments respectively. However, intracytoplasmic fibrils are also found in granular amoebocytes containing Type A granules (Fig. 4). Thus, it is conceiv-
able that granular amoebocytes could also be the precursor of the
fibroblastlike cells. While the association of fibroblastlike
cells with encapsulation and wound repair is well documented,
critical experiments designed to reveal the potential of agranular
amoebocytes to transform into fibroblastlike cells and myoblasts,
have yet to be performed. Tripp *et al.* (1966) note that fibroblastlike cells are occasionally seen in blood cell preparations.
Feng and Feng (1974) show that the mobilization of fibroblastlike
cells is extremely rapid, these cells could make their appearance
as early as 1 hr post-injection of avian erythrocytes into
C. virginica. This observation coupled with the presence of
fibroblastlike cells in the control as well as experimental blood
smears, strongly suggests that this cell type is a part of the
normal formed hemolymph elements. In light of these data, one

Table 3. Comparison of the size variation of cytoplasmic granules in the amoebocytes of *Grassostrea virginica* and *Crassostrea gigas*.

Type of Granules	*Crassostrea virginica*	*Crassostrea gigas*	
A	0.82 ± 0.16[1]	0.86 ± 0.27[2]	0.7-1.2[3]
B	0.59 ± 0.16	0.60 ± 0.12	-------
C	0.85 ± 0.16	-------	-------
C'	-------	0.34 ± 0.13	0.4-0.56

The figures represent mean diameter of granules in μ ± 1 S.D.
[1]From Feng *et al.* (1971); [2]From this study; [3]From Ruddell (1971b).

Type A   cv vs. cg: t = 1.697, d.f. = 502, P >0.10.
Type B   cv vs. cg: t = 0.031, d.f. = 263, P >0.50.

must assume that under normal conditions circulating fibroblastlike cells are probably being renewed by transformation from the
agranular and granular amoebocytes. In an emergency, the
immediate need for such cell type in defense could be quickly
fulfilled by assembling all available circulating fibroblastlike
cells to the site of inflammation, and then followed by a slower
process of transformation of amoebocytes into the fibroblastlike
cells, which serves to replenish and reinforce those cells
mobilized earlier. Thus, the available evidence seems to favor
the view that mobilization of this cells type is biphasic in the
oysters.

Several observations of our study do not appear to be in
accord with that of Ruddell (1971a,b). He states that "the
granulocytes were almost never observed phagocytosing particulate
material", and that "carmine, carbon particles, and bacteria
injected in seawater suspensions into oyster mantle were phato-
cytosed exclusively by the granular amoebocytes". In contrast to
his findings, we have seen both agranular and granular amoebocytes
of *C. virginica* and *C. gigas* participating in phagocytosis of a
variety of foreign particles (Feng *et al.*, 1971; Figs. 7,8). The
discrepancies may have been due to the source, the physiological
state of the amoebocytes involved, and the experimental procedure
employed in the two studies.

Ruddell (1971c) also reports that acid phosphatase was not
present in the acidophilic granules; it was also absent in
lysosomes of *Mercenaria mercenaria* (Janoff and Howrylko, 1964).
Contrary to these statements, we have demonstrated the presence
of granule-associated acid and alkaline phosphatase, and non-
specific esterases in the granular amoebocytes of *C. virginica*
by a modified method of Burstone (1957, 1958) (Feng *et al.*, 1971)
and acid phosphatase reaction products in the phagosome of granular amoebocytes in *C. gigas* (Fig. 8). Our positive results have
also been shared by Eble (1966) and Eble and Tripp (1968). These
apparent inconsistencies, particularly in the case where the
same species of oysters is used, can be ascribed to the techniques
used in demonstrating the above mentioned intracellular enzymes.
In our experience, most enzyme histochemical procedures devised
with mammalian tissues in mind, do not work with invertebrate
tissues; appropriate modifications are often required. Our
success in demonstrating the above mentioned hydrolytic enzymes
in the oyster amoebocytes, is attributable to the fact that
fixation was eliminated. We have found that oyster intracellular
hydrolytic enzymes are extremely sensitive to fixatives such as
formaldehyde, methanol, and absolute ethanol at room temperatures.
Such sensitivity to fixatives probably exists in a number of
molluscan tissues. More importantly, these observations probably
reflect the basic difference of mammalian and invertebrate
intracellular enzyme systems.

Ruddell (1971b) states in rather unequivocal terms that "the granular amoebocytes were <u>never</u> observed under going mitosis" and that "replication was confined to the agranular amoebocytes, which were often observed dividing." Our results presented herein do not agree with the above statements. Binucleated granular amoebocytes (Fig. 3) and amoebocytes containing centriole are encountered fairly frequently in both Giemsa stained as well as EM preparations. In our view, Ruddell's observations may represent a condition peculiar to wound repair and are probably exceptions rather than the rule.

Morphology and function of amoebocytes of *Mytilus* are not very well known. Wharton (1846) recognized three types of amoebocytes in the hemolymph of *Mytilus* and suggested that they represent developmental stages (Takatsuki, 1934). White (1937) stated that *Mytilus* amoebocytes were completely filled with granules. In our studies, several types of formed elements are distinguishable in the hemolymph of *Mytilus coruscus*. Agranular amoebocytes are relatively simple in structure. There are three types of granular amoebocytes probably reflecting their varied functions. Among these the amoebocytes with mixed types of granules are most distinct (Fig. 10). One can discern at least four types of granules (Type A,B,C, and D) in these cells, but the question whether these granules represent developmental stages or are truly polymorphic, is yet to be determined.

Phagocytosis of carmine and bacteria has been shown in *Mytilus edulis* (Mikhailova and Prozdnikov, 1961, 1962). Results from the acid phosphatase studies suggest that Type C granules are probably instrumental in the destruction of phagocytosed bacteria (Fig. 13).

Although carotenoids have been reported from *Mytilus edulis* (Fisher *et al.*, 1956), *M. californianus* (Campbell, 1970) and other bivalve molluscs: *Pectunculus glycymeris* (Lederer, 1933; Fabre and Lederer, 1934), *Pecten maximus* (Lederer, 1934), the pigments are usually extracted from the whole animal. In this study, the presence of flavonoids, β-carotene, canthaxanthin, and other unidentified xanthophylls has been demonstrated in the amoebocytes of *M. coruscus* and *C. gigas*. It is probable that these pigments are associated with the lipid like bodies in the amoebocytes of the two species of pelecypods.

There is no doubt that the two species of lamellibranchs acquire carotenoids from unicellular algae, their diet, since both are known filter feeders. However, it is significant to note the presence of carotenoids in amoebocytes bled from the adductor muscle sinus, which strongly supports the contention that food-particle-laden amoebocytes in the lumen of intestine traverse the intestinal epithelium into the deep tissues (Yonge, 1926). Based on the results of Yonge (1926), Stauber (1950), Tripp (1960), and the present study, we can conclude now that there is a two-way traffic of lamellibranch amoebocytes across the intestinal epithelium and that the movement of amoebocytes is clearly not a

random phenomenon. The basic question here is to answer what are
the mechanisms that dictate the direction of amoebocyte movement
from the lumen to deep tissues and vice versa. Such directional
movement of amoebocytes is probably different from that induced
by chemotaxis agents although experimental evidence is still
lacking. One may infer that the amoebocytes possess a highly
developed ability to distinguish the food particles from non-
destructive particles: spores of bacteria or India ink administered
via the parental route, which in turn influence the direction of
movement of the amoebocytes.

Feng *et al.* (1971) have reported the presence of intragranular glycogen in Type C granules of *C. virginica* amoebocytes.
In a recent study, Cheng and Cali (1974) suggest that glycogen may
be synthesized by the granular amoebocyte of *Crassostrea virginica*
from breakdown products of phagocytosed bacteria. It would be
interesting to verify the ultrastructural observations of these
investigators by introducing C^{14} labelled bacteria or algae
to the amoebocytes and subsequently determining the hot glycogen
content of the amoebocytes.

Verne (1926) observed that pigment fragment present in amoebocytes were irregular in shape which was interpreted as being
characteristic of excretory chromatic materials. Storage granules,
on the other hand, tend to be more structured. Pathological pigments, such as bile pigments (Jirsa, 1969), lipofucsins (Porta
and Hartroff, 1969), and heavy metal compounds (Gedligk, 1969),
have been shown to be stored in lysosomelike bodies and phagosomes
which imply scavenging of the pigments by amoebocytes. Ceroid
bodies and brown cells are known to be associated with pathological
conditions in *C. virginica* (Mackin, 1951, 1962; Stein and Mackin,
1955; Cheng and Burton, 1965).

Flavonoids are chain-linked benz-chroman derivatives which
form several very large classes of plant chrome; they are the
major contributors of flower chromes. It has been documented
that flavonoids are present in larval and adult Lepidoptera and
in herbivorous Hemiptera and Coleoptera. These compounds have
also been reported in *Helix*, the garden snail (Kubista, 1950),
and *Sertularella*, the marine hydroid (Payne, 1931). The source of
flavonoids in the amoebocytes of *C. gigas* and *M. coruscus* is
probably also of dietary origin. However, the biochemical significance of flavonoids in amoebocytes remains to be investigated.
The possibility of these compounds serving as electron transport
systems is particularly intriguing, since most flavonoids possess
redox properties.

V. ACKNOWLEDGEMENT

We thank Professor Masao Yoshida, Director of Tamano Marine
Laboratory, Okayama University, for his valuable suggestions as
well as providing field and laboratory facilities during the
course of this study.

We are also indebted to Mr. M. Isozaki for collecting the specimens and preparing of electron micrographs, to Mr. I. Yoshii for his skillful assistance in TLC, and to Mr. H. Ohtsuki of the Second Physiology Department, Medical School, Okayama University, for his technical assistance in the use of special equipment. The first and second authors are most appreciative for the hospitality so graciously bestowed upon them by Professor Yoshida and his staff during their stay at the Laboratory.

SUMMARY

An ultrastructural study of the circulating amoebocytes of *Crassostrea gigas* and *Mytilus edulis* has revealed two basic cell types: agranular and granular amoebocytes. Based upon their morphological features as well as the texture of intragranular contents, several types of cytoplasmic granules are recognized. The functional significance of amoebocytes in defense is demonstrated by their ability to phagocytose and degrade *Escherichia coli*. Acid phosphatase reaction products have been localized in Golgi cisternae, certain granules, and phagosomes which contain ingested bacteria.

The role in nutrition played by the amoebocytes is shown by using flavonoids, β-carotene, canthaxanthin, and other unidentified xanthophylls, which are known to be of plant origin, as natural tracers; these compounds are revealed by the use of TLC in amoebocytes which were bled from the adductor muscle sinus. This observation confirms the role of these cells in transporting food particles across the intestinal epithelium as originally suggested by Yonge (1926) in *Ostrea edulis*. Our data, when considered with the findings of Stauber (1950), Tripp (1960), and Cheng and Cali (1974), suggest that the directional movement of particle-laden amoebocytes from the intestinal lumen to deep tissues and vice versa, can not be explained on the basis of recognition of self or nonself. Attrition of particle-laden amoebocytes as a defense mechanism, is discussed in the context of digestive process as a whole.

REFFERENCES

Burstone, M.S.(1957). The cytochemical localization of esterase. *J. Nat. Cancer Inst.*, 18, 167-172.
Campbell, S. A.(1970). The carotenoid pigments of *Mytilus edulis* and *Mytilus californianus*. *Comp. Biochem. Physiol.*, 32, 97-115.
Cheng, T. C.(1967). Marine molluscs as hosts for symbioses with a review of known parasites of commercially important species. *In*: Advanced Marine Biology, Vol. 5, 1-424, F. S. Russell, Ed., Academic Press, London.

Cheng, T. C. and Burton, R. W. (1965). Relationships between
    Bucephalus sp. and *Crassostrea virginica:* Histopathology
    and site of infection. *Chesapeake Sci.*, 6, 3-16.
Cheng, T. C. and Cali, A. (1974). An electron microscope study of
    the fate of bacteria phagocytosed by granulocytes of
    Crassostrea virginica. *In:* Contemporary Topics in
    Immunobiology, Vol. 4 Invertebrate Immunology, 25-35p.,
    E. L. Cooper, Ed., Plenum Press, New York and London.
Cheng, T. C. and Rodrick, G. E. (1974). Identification and
    characterization of lysozyme from the hemolymph of the soft-
    shelled clam, *Mya arenaria.* *Biol. Bull.*, 147, 311-320.
Cheng, T. C., Rodrick, G. E., Foley, D. A., and Koehler, S. A.
    (1975). Release of lysozyme from hemolymph cells of
    Mercenaria mercenaira during phagocytosis. *J. Invertebr.*
    Pathol., 25, 261-265.
Eble, A. F. (1966). Some observations on the seasonal distribution
    of selected enzymes in the American oyster as revealed by
    enzyme histochemistry. *Proc. Nat. Shellfish. Assoc.*, 56,
    37-42.
Eble, A. F. and Tripp, M. R. (1968). Enzyme histochemistry of
    phagosomes in oyster leucocytes. *Bull. N. J. Acad. Sci.*,
    B 13, 93.
Fabre, R. and Lederer, E. (1934). Contributions à l'étude des
    lipochromes des animaux. *Bull. Soc. Chim. Biol.*, 16, 105-118.
Farquhar, M. G., Bainton, D. F., Baggiolini, M. and deDuve, C.
    (1972). Cytochemical localization of acid phosphatase
    activity in granule fractions from rabbit polymorphonuclear
    leukocytes. *J. Cell Biol.*, 54, 141-156.
Feng, S. Y. (1967). Responses of molluscs to foreign bodies, with
    special reference to the oyster. *Fed. Proc.* 26, 1685-1692.
Feng, S. Y. and Feng, J. S. (1974). The effect of temperature on
    cellular reactions of *Crassostrea virginica* to the injection
    of avian erythrocytes. *J. Invertebr. Pathol.*, 23, 22-37.
Feng, S. Y., Feng, J. S., Burke, C. N. and Khairallah, L. H.
    (1971). Light and electron microscopy of the leucocytes of
    Crassostrea virginica (Mollusca: Pelecypoda). *Z. Zellforsch.*,
    120, 222-245.
Fisher, L. R., Kon, S. K. and Thompson, S. Y. (1956). Vitamin A
    and carotenoids in certain invertebrates-IV. Mollusca:
    Loricata, Lamellibranchiata and Gastropods. *J. Mar. Biol.*
    Assoc. U. K., 35, 41-61.
Foley, D. A. and Cheng, T. C. (1972). Interaction of molluscs and
    foreign substances: the morphology and behavior of hemolymph
    cells of the American oyster, *Crassostrea virginica, in vitro.*
    J. Invertebr. Pathol., 19, 383-394.
Gedigh, P. (1969). Pigmentation caused by inorganic material. *In:*
    Pigments in Pathology, pp. 1-32. M. Wolman, Ed., Academic
    Press, New York.
George, W. C. (1952). The digestion and absorption of fat in
    lamellibranchs. *Biol. Bull.*, 102, 118-127.

Janoff, A. and Hawrylko, E. (1963). Lysosomal enzymes in invertebrate leucocytes. *J. Cell. Comp. Physiol.*, 63, 267-271.

Jirsa, M. (1969). The bile pigments. *In:* Pigments in pathology, 151-190 p. M. Wolman, Ed., Academic Press, New York-London.

Kubista, V. (1950). Flavones in *Helix pomatia. Experientia,* 6, 100.

Lederer, E. (1933). Note sur un noveau caroténoide trouvé dans le Pétoncle *(Pectunculus glycymeris* L.). *C. R. Séanc. Soc. Biol.,* 113, 1015-1016.

Lederer, E. (1934). Sur un nouveau caroténoide trouvé dans la Coquille Saint-Jacques *(Pecten maximus)*. *C. R. Séanc. Soc. Biol.* 116, 150-153.

Mackin, J. G. (1951). Histopathology of infections of *Crassostrea virginica* (Gmelin) by *Dermocystidium marinum* Mackin, Owen and Collier. *Bull. Mar. Sci. Gulf Carib.,* 1, 72-87.

Mackin, J. G. (1962). Oyster disease caused by *Dermocystidium marinum* and other microorganisms in Louisiana. *Publ. Inst. Mar. Sci.* Univ. Texas, 7, 132-229.

Mikhailova, I. G. and Prazdnikov, E. V. (1961). Two questions on the morphological reactivity of mantle tissues in *Mytilus edulis* L. *Tr. Murmansk. Morsk. Biol. Inst.,* 3, 125-130.

Mikhailova, I. G. and Prazdnikov, E. V. (1962). Inflammatory reactions in the Barents Sea sea mussel *(Mytilus edulis* L.). *Tr. Murmansk. Morsk. Biol. Inst.,* 4, 208-220.

Narain, A. S. (1972). Blood chemistry of *Lamellidens corrianus. Experientia,* 28, 507.

Nelson, T. C. (1933). On the digestion of animal forms by the oyster. *Proc. Soc. Exptl. Biol. Med.,* 30, 1287-1290.

Ohuye, T. (1937). On the coelomic corpuscles in the body fluid of some invertebrates. VII. On the formed elements in the body fluid of some marine invertebrates which possess the red blood corpuscles. *Sci. Rep. Tohoku Imp. Univ.,* Ser. 4, 12, 593-622.

Ohuye, T. (1938b). On corpuscles in the body fluids of some invertebrates. XII. General consideration of the results obtained by preceding investigations. *Sci. Rep. Tohoku Imp. Univ.,* Ser. 4, 13, 359-380.

Owen, G. (1966). Digestion. *In:* Physiology of Mollusca, Vol. 2, pp. 53-96. K. M. Wilbur and C. M Yonge, Eds., Academic Press, New York.

Payne, N. M. (1931). Hydroid pigments I: General discussion and pigments of the Sertulariidae. *J. Mar. Biol. Assoc., U.K.,* 17, 739-749.

Porta, E. A., and Hartroft, W. S. (1969). Lipid pigment in relation to aging and dietary factors. *In:* Pigments in Pathology, M. Wolman, Ed. pp. 191-235. Academic Press, New York-London.

Potts, W.T.W. (1967). Excretion in the molluscs. *Biol. Rev.,* 42, 1-41.

Purchon, R. D. (1968). The Biology of the Mollusca. Permamon Press, London.

Rifkin, E., Cheng, T. C. and Hohl, H. R. (1969). An electron-microscope study of the constituents of encapsulating cysts in the American oyster, *Crassostrea virginica*, formed in response to *Tylocephalum metacestodes*. *J. Invertebr. Pathol.*, 14, 211-226.

Ruddell, C. L. (1971a). The fine structure of oyster agranular amebocytes from regenerating mantle wounds in the Pacific oyster, *Crassostrea gigas*. *J. Invertebr. Pathol.*, 18, 260-268.

Ruddell, C. L. (1971b). The fine structure of the granular amebocytes of the Pacific oyster, *Crassostrea gigas*. *J. Invertebr. Pathol.*, 18, 269-275.

Ruddell, C. L. (1971c). Elucidation of the nature and function of the granular oyster amebocytes through histochemical studies of normal and traumatized oyster tissues. *Histochemie*, 26, 98-112.

Stauber, L. A. (1961). Immunity of invertebrates, with special reference to the oyster. *Proc. Nat. Shellfish. Assoc.*, 50, 7-20.

Stein, J. E. and Mackin, J. G. (1955). A study of the nature of pigment cells of oysters and the relation of their numbers to the fungus disease caused by *Dermocystidium marinum*. *Texas J. Sci.*, 7, 422-429.

Stevenson, R. N. and South, G. R. (1975). Observations on phagocytosis of *Coccomyxa parasitica* (Coccomyxaceae; Chlorococcales) in *Placopecten magellanicus*. *J. Invertebr. Pathol.*, 25, 307-311.

Takatsuki, S. (1934). On the nature and functions of the amoebocytes of *Ostrea edulis*. *Quart. J. Microsc. Sci.*, 76, 379-431.

Tripp, M. R. (1960). Mechanisms of removal of injected microorganisms from the American oyster, *Crassostrea virginica* (Gmelin). *Biol. Bull.*, 199, 273-282.

Tripp, M. R. (1963). Cellular responses of mollusks. *Ann. N. Y. Acad. Sci.*, 113, 467-474.

Tripp, M. R. (1970). Defense mechanisms of Molluscs. *J. Reticuloendothel. Soc.*, 7, 173-182.

Tripp, M. R., Bisignani, L. A. and Kenny, M. T. (1966). Oyster amoebocytes in vitro. *J. Invertebr. Pathol.*, 8, 137-140.

Verne, J. (1926). Les Pigments dans L'Organisme Animale. Gaston Doin et Cie, Paris.

Wagge, L. E. (1955). Amoebocytes. *Int. Rev. Cytol.*, 4, 31-78.

Wharton, J. T. (1846). The blood corpuscle considered in its different phases of development in the animal series. Memoir II. Invertebrata. *Phil. Trans. Roy. Soc.*, 136, 89-102.

White, K. M. (1937). Mytilus. *Liverp. Mar. Biol. Comm. Mem.*, 31, 1-117.

Wilbur, K. M. (1964). Shell formation and regeneration. *In:* Physiology of Mollusca, Vol. 1, pp. 243-282, K. M. Wilbur and C. M. Yone, Eds. Academic Press, New York and London.

Yonge, C. M. (1926). Structure and physiology of the organs of feeding and digestion in *Ostrea edulis*. *J. Marine Biol. Ass. U.K.,* 14, 295-386.

Yonge, C. M. (1937). Evolution and adaptation in the digestive system of metazoa. *Biol. Rev.,* 12, 87-115.

Yonge, C. M. (1946). Digestion of animals by lamellibranchs. *Nature,* 157, 729.

Zacks, S. I. (1955). The cytochemistry of the amoebocytes and intestinal epithelium of *Venus mercenaria* (Lamellibranchiata) with remarks on a pigment resembling ceroid. *Quart. J. Microsc. Sci.,* 96, 57-71.

Zacks, S. I. and Welsh, J. H. (1953). Cholinesterase and lipase in the amoebocytes, intestinal epithelium and heart muscle of the quahog, *Venus mercenaria*. *Biol. Bull.,* 105, 200-211.

Immune Responses in the Earthworm, Aporrectodea trapezoides (Annelida), Against Rhabditis pellio (Nematoda)

GEORGE O. POINAR, JR.

AND

ROBERTA T. HESS

DIVISION OF ENTOMOLOGY AND PARASITOLOGY
UNIVERSITY OF CALIFORNIA
BERKELEY, CALIFORNIA

I. INTRODUCTION

Immune responses of earthworms to nematodes or other metazoan parasites have received little attention. While investigating the behavior of *Rhabditis pellio* in the earthworm, *Aporrectodea trapezoides*, Poinar and Thomas (1976) noted that most nematodes occurred in the bladders and tubules of the metanephridia. However, those that entered the coelom were promptly encapsulated and incorporated into "brown bodies." These "brown bodies" not only contained nematodes but also protozoan cysts and other debris that occurred in the worm's coelom. It was clear that the "brown bodies" constituted an immune response on the part of the worm which is probably typical for oligochaete annelids. The purpose of this investigation was to examine the ultrastructure of the "brown bodies" of *A. trapezoides* containing *R. pellio*.

II. MATERIALS AND METHODS

Specimens of the earthworm, *Aporrectodea trapezoides*, infected with the nematode *Rhabditis pellio* were obtained from Riverside, California.

"Brown bodies" containing *R. pellio* were obtained by opening the terminal segment of the earthworm and squeezing out the coelomic fluid. For electron microscopy, the "brown bodies" were fixed in 4.0% glutaraldehyde in Millonig's phosphate buffer for 1 hr, then transferred to a 1% solution of osmium tetroxide for 1 hr at 4°C. Following fixation, the "brown bodies" were rinsed in buffer, dehydrated in an alcoholic series, and embedded in Araldite 6005. Sections made with glass knives mounted in a Porter-Blum MT-2 microtome were stained with saturated aqueous uranyl acetate followed by lead citrate and examined with an RCA-3F and a Philips EM-300 electron microscope.

III. RESULTS

The "brown bodies" of *A. trapezoides* containing juvenile and young adults of *R. pellio* occurred throughout the earthworm's coelom, although most were concentrated in the posterior few segments. The smaller and presumably newer formed "bodies" were yellow or light brown, while the larger ones were dark brown. The "brown bodies" were roughly elliptical in shape with most measuring 1.0 by 0.5 mm, although some reached 2.0 mm in length and 1.0 mm in width (Fig. 1).

The nematodes appeared to be randomly distributed in the "brown bodies." Most were living and in a partially coiled state with movement restricted to a slight twitching of both extremities (Fig. 2). Dead nematodes were in various stages of degeneration. Once enclosed in a "brown body," it is highly unlikely that the nematodes could ever escape into the coelom again.

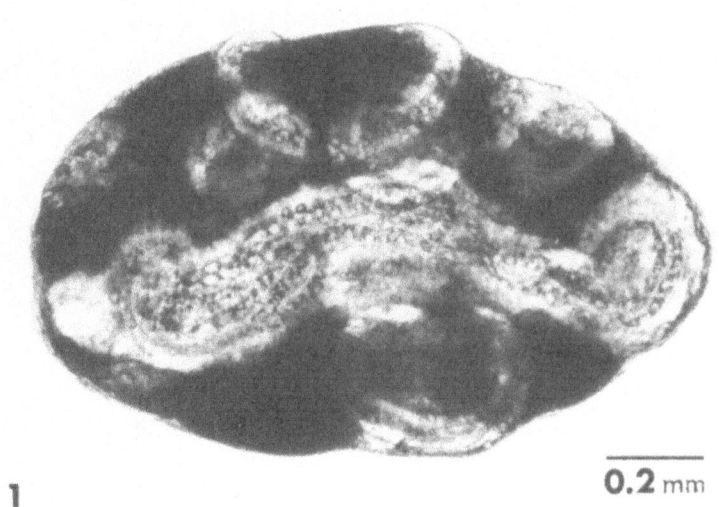

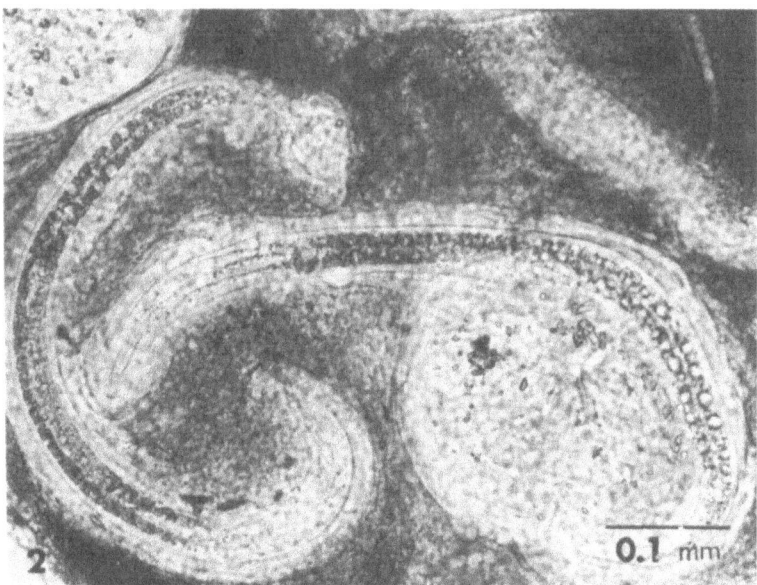

Fig. 1. A "brown body" from *Aporrectodea trapezoides* containing
         Rhabditis pellio.

Fig. 2. Two specimens of *Rhabditis pellio* coiled in a "brown body"
         of *Aporrectodea trapezoides*. Note lighter uniform deposit
         surrounding the nematodes.

Fig. 3. One type of amoebocytic coelomocyte of *Aporrectodea trapezoides* involved in "brown body" formation. Note the small, elliptical electron-dense granules (G) and the Golgi (GA). Bar equal 1 μm.

Fig. 4. A second type of amoebocytic coelomocyte of *Aporrectodea trapezoides* involved in "brown body" formation. Note the small elliptical granules (G) and large dense granules (LG). L = Lipid droplets. Bar equals 1 μm.

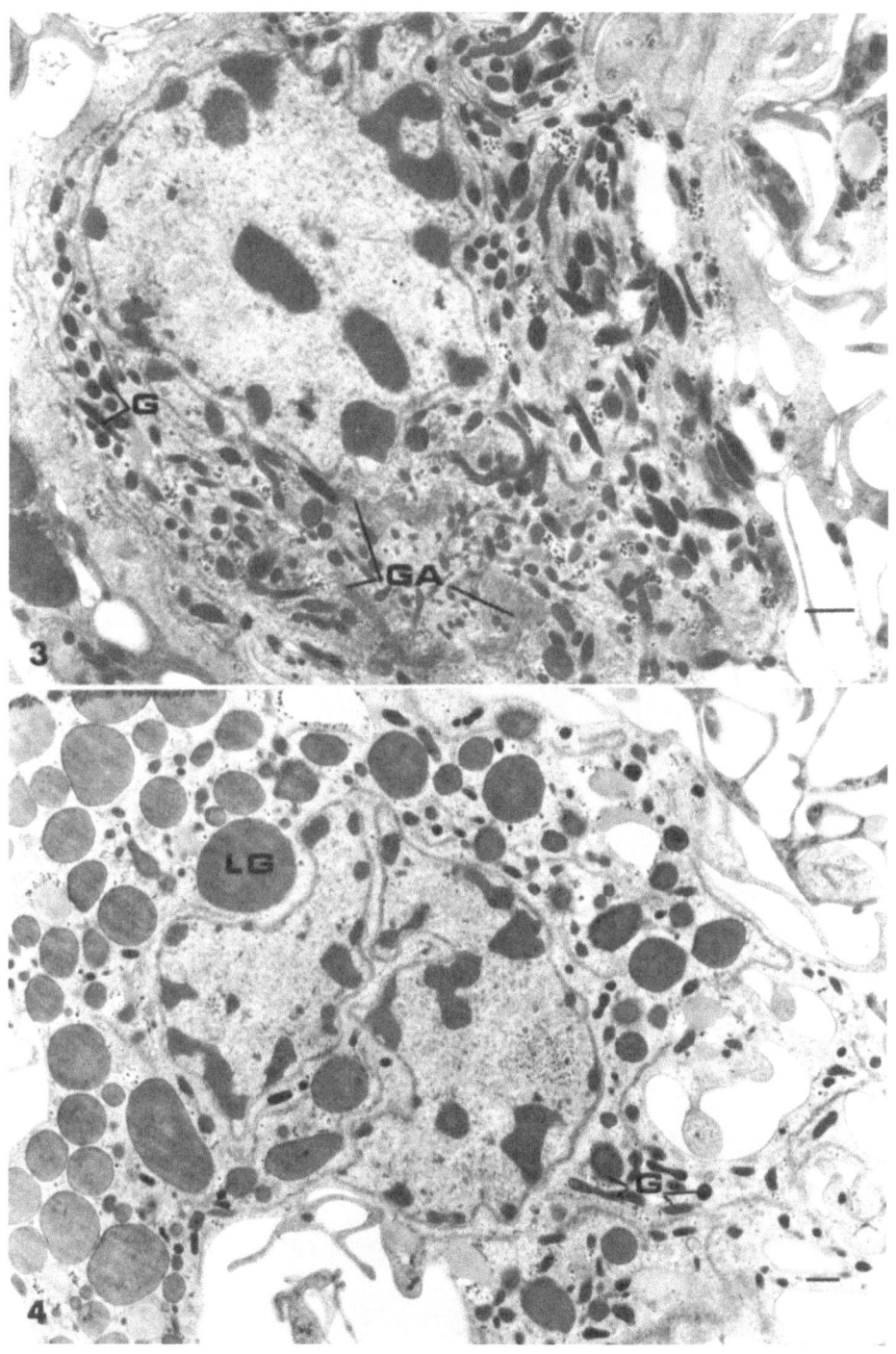

Fig. 5.   Section through the homogeneous deposit (H) surrounding
          Rhabditis pellio (N) in a "brown body" of *Aporrectodea
          trapezoides*.   B = region of amoebocytes.

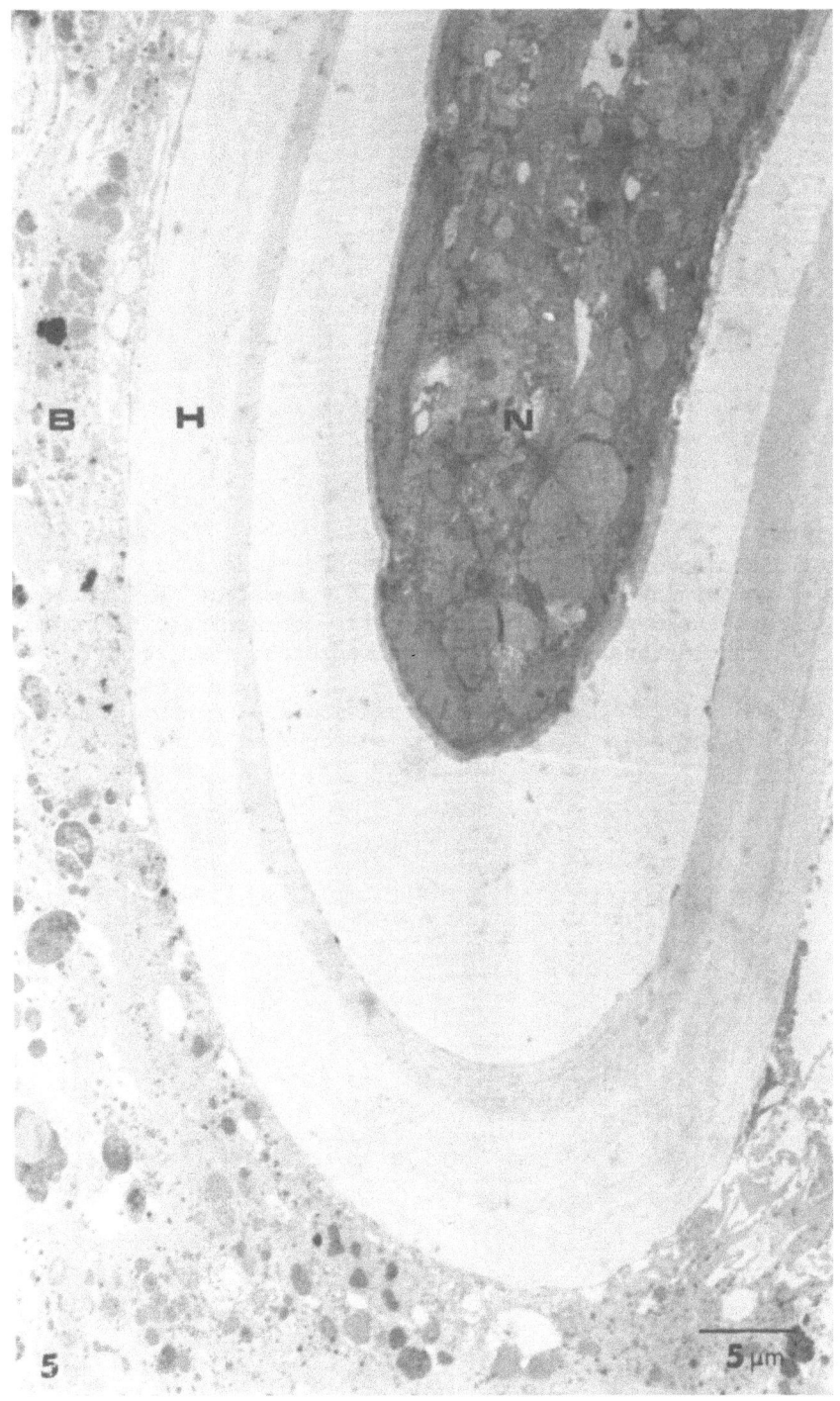

Fig. 6. Section of a "brown body" of *Aporrectodea trapezoides*
         containing *Rhabditis pellio*. N = nematode H = homo-
         geneous deposit B = blood cells containing pleomorphic
         granules bearing electron-dense zones. Bar equals 1 μm.

Fig. 7. Amoebocyte of *Aporrectodea trapezoides* in contact with
         the homogeneous deposit (H) surrounding *Rhabditis
         pellio*. Bar equals 1 μm.

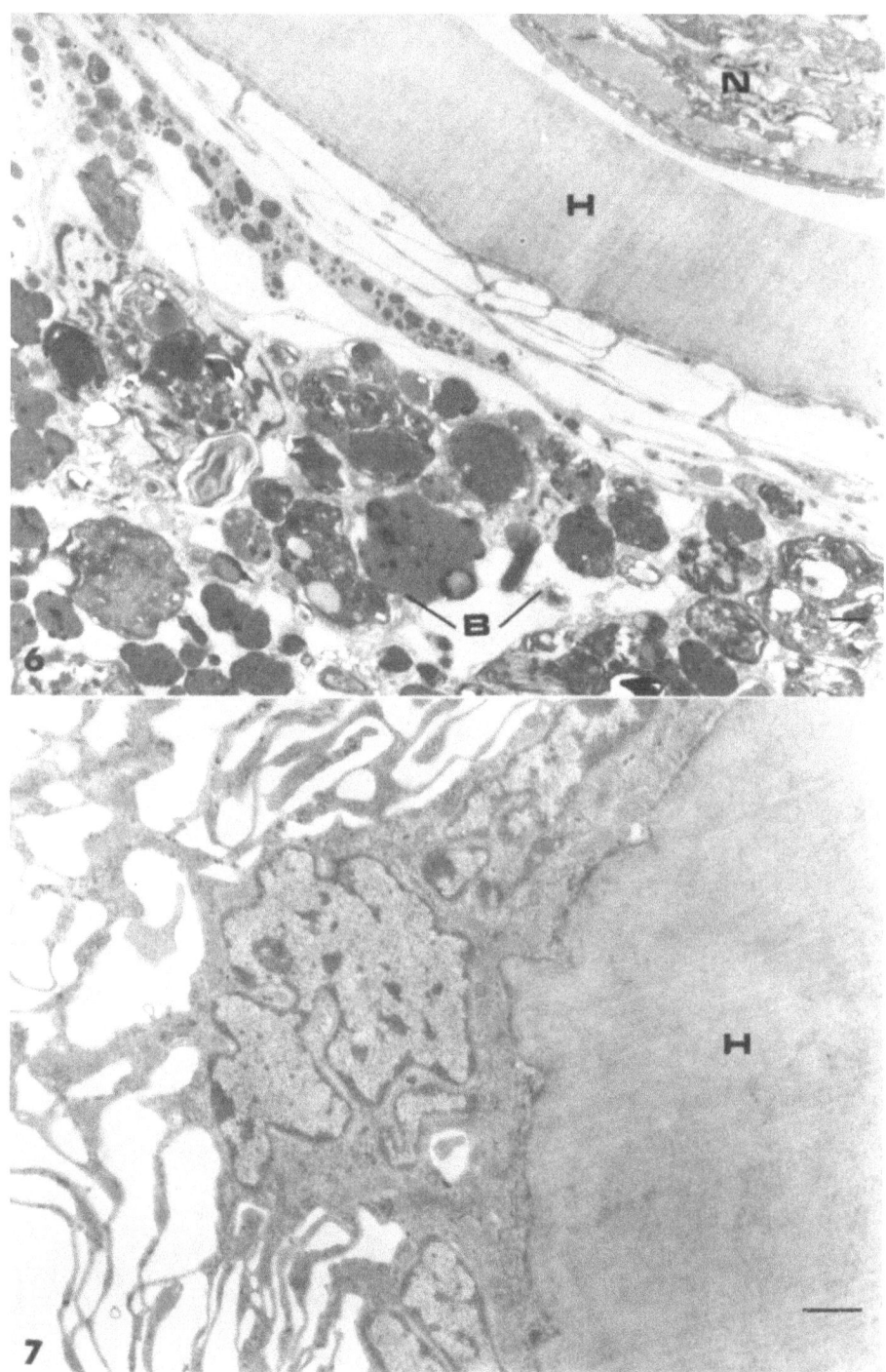

Fig. 8. Section through amoebocytes comprising the cellular
         portion of the "brown bodies" of *Aporrectodea trapezoides*.
         Note extensive branching of the pseudopodia. Bar equals
         1 μm.

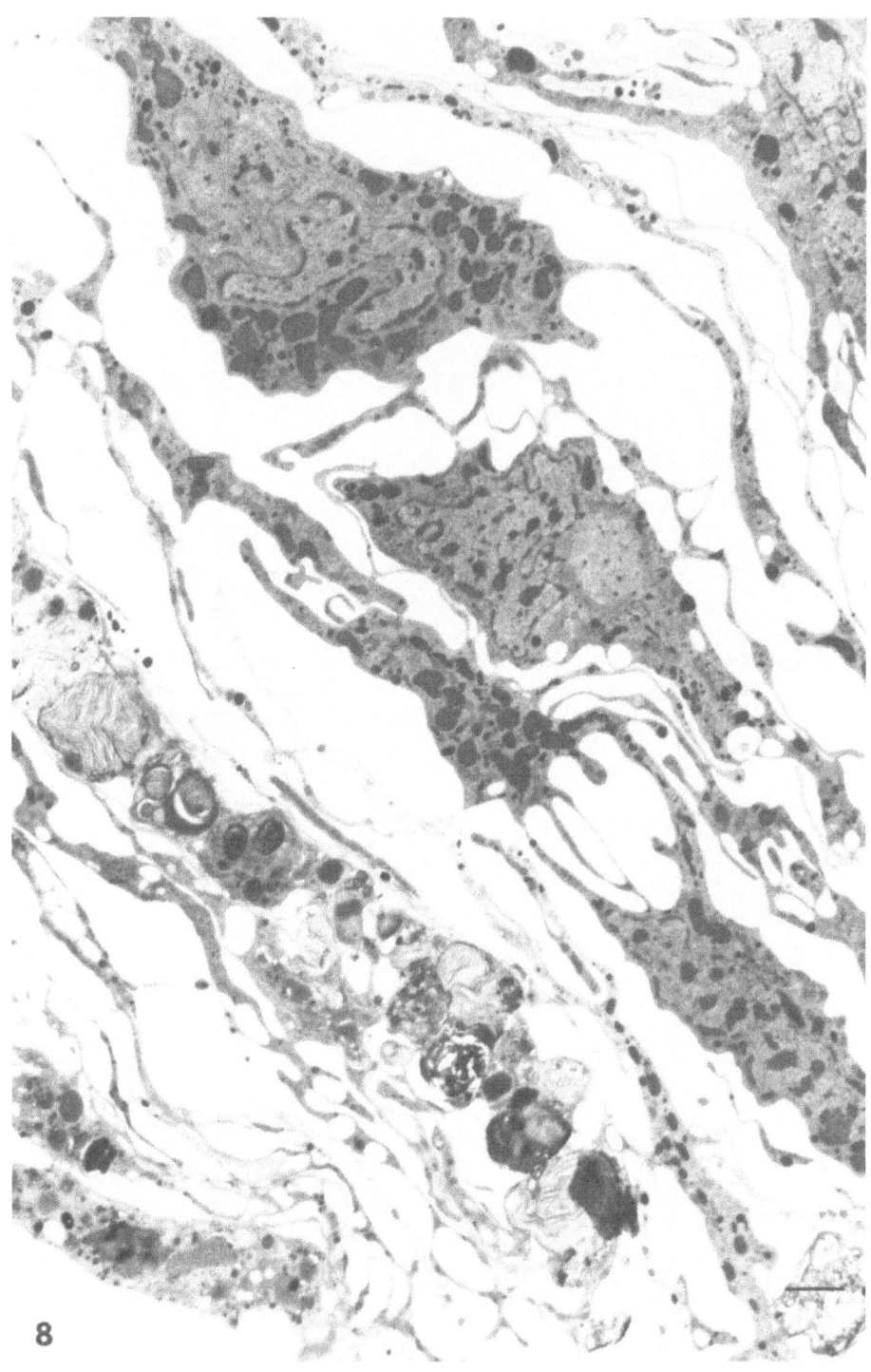

8

Fig. 9. Darkening of the homogeneous deposit surrounding
          Rhabditis pellio (N) in a "brown body" of *Aporrectodea*
          trapezoides. Bar equals 1 μm.

Fig. 10. Impressions made in the dark homogeneous deposit from
          the annules of the adjacent cuticle of *Rhabditis pellio*
          (N) in a "brown body" of *Aporrectodea trapezoides*.

Electron microscope observations showed that the "brown bodies"
consisted of host amoebocytic coelomocytes. Several different
morphological types of amoebocytes were involved in the response.
One type contained small elliptical electron-dense granules which
appeared to originate from the Golgi bodies (Fig. 3). A second
type contained large spherical granules in addition to the small
elliptical granules (Fig. 4). A third type contained large
pleomorphic granules bearing electron-dense zones (Figs. 5, 6).
This latter type often contained variable amounts of phagocytized
and degenerating material (Fig. 6). The various types of amoebocytes probably represent different phases or physiological states
of the same basic coelomocyte.

Adjacent to the nematode and separating the inner layer of
amoebocytes from the nematode's cuticle was a uniform homogeneous
deposit (Figs. 2, 5). This deposit was generally about 7.5 µm
(2.0-12.0 µm) thick and completely surrounded the nematodes. A
similar deposit surrounded protozoan cysts and chaetae that were
also incorporated into "brown bodies." No evidence of cells or
cellular debris occurred within this deposit. At high magnification, the deposit showed a faint layering effect suggesting
that it was formed over a period of time (Figs. 5, 6). The
initial layer of amoebocytes surrounding this deposit lacked
inclusion granules and phagocytized material (Figs. 6-8).

The unique feature of the amoebocytes making up the bulk of the
"brown bodies" was their long branching pseudopodia (Fig. 8).
The pseudopodia of adjacent amoebocytes interlocked with obvious
junctions, but the intricate patterns of branching made it
difficult to definitely establish whether cellular fusion occurred.

In large and presumably older "brown bodies," the homogeneous
deposit surrounding the nematode frequently became electron dense
and rigid (Figs. 9, 10). The inner edge of the deposit often
contained impressions made from contact with the nematodes cuticle
(Fig. 10). In these "brown bodies," many of the amoebocytes near
the nematode showed signs of degeneration.

IV. DISCUSSION

It is now clear that the "brown bodies" of *A. trapezoides,* and
probably other earthworms, are actually single or multiple
capsules formed primarily by host amoebocytes.

Nematodes and other foreign objects in the capsules were
surrounded by a homogeneous deposit. Metchnikoff (1892) described
this deposit as a supplementary cuticle of the enclosed nematode;
however, this would not explain the deposit around other objects,
and ultrastructural features do not support this idea. Cuénot (1898)
considered the deposit as a secretion of the host blood cells.
However, it is difficult to see how such a thick, uniform,
homogeneous deposit could arise from individual blood cells. The
cells would have to release their deposit and then retreat

for others to do the same. Also, there was no indication of cellular debris in the deposit which was approximately the same thickness on newly formed capsules with few attached amoebocytes as on older capsules containing many blood cells.

Our conclusions are that the homogeneous deposit is a humoral response formed by the noncellular portion of the blood coagulating on foreign objects as soon as they enter the coelom. This response is similar to the initial response of *Culex pipiens* larvae to invading nematodes (Poinar and Leutenegger, 1971). Within 1 hr after entering the mosquito larvae, a homogeneous deposit originating from components in the noncellular portion of the hemolymph surrounded the nematode.

In the present investigation, the homogeneous deposit eventually darkened and became rigid, suggesting a melanization reaction. Although tests to determine the presence of melanin were not made, the nature of the material certainly resembled melanin found in insect hosts attacked by nematodes (Poinar and Leutenegger, 1971).

In *A. trapezoides*, the initial deposit was surrounded by layers of host amoebocytes that, together with other foreign objects, formed huge multiple capsules. The later darkening reaction served to further entrap the nematode and release was probably impossible. This may be why cultures of *R. pellio* were rarely established by placing "brown bodies" containing living nematodes on nutrient agar (Poinar and Thomas, 1976).

The type of blood cell involved in the encapsulation reaction was the phagocytic amoebocyte or basic cell type of oligochetes (Avel, 1959). According to Liebmann (1942), the amoebocytes are leucocytes containing flattened or reniform nuclei and pseudopodia and are known to play a role in wound healing, phagocytosis, and encapsulation. Tu (1937) described the "brown bodies" of earthworms as composed of circulating or lymph cells and chloragogen cells. Liebmann (1942) also stated that choragogen cells are incorporated into "brown bodies." In the present study, however, no chloragogen cells were encountered during the ultrastructural investigation of the "brown bodies."

"Brown body" or capsule formation in *A. trapezoides* is a defense reaction which in the case of *R. pellio* is highly successful. Living, dying, and dead nematodes were all encountered in the "brown bodies" removed from the earthworm's colon.

Thus, it is obvious that living nematodes are attacked and killed by this host response, which is actually twofold. The first response is humoral and consists of a homogeneous deposit from the noncellular portion of the blood on the nematode. This is followed by an encapsulation reaction where host amoebocytes surround the nematode and its original homogeneous deposit. These two responses result in an effective defense against *R. pellio* and probably are typical of oligochaete annelids.

V. ACKNOWLEDGMENTS

The authors wish to thank Drs. H. H. Shorey and Jo-Ann S. Gaston of the Department of Entomology, University of California at Riverside, for generously supplying infected earthworms and G. E. Gates for identifying the worms.

SUMMARY

The present investigation examines the ultrastructure of the "brown bodies" of the earthworm, *Aporrectodea trapezoides*, enclosing the nematode *Rhabditis pellio*. The nematodes were initially surrounded by a uniform homogeneous deposit thought to originate from the noncellular portion of the blood and thus representing a humoral response. This homogeneous deposit was then surrounded by host amoebocytes that made up the remainder of the "brown bodies."

The "brown bodies", therefore, are host capsules representing a defense response which is highly successful in the case of *R. pellio*. Escape of encapsulated nematodes seemed highly unlikely, and many were eventually destroyed.

REFERENCES

Avel, M. 1959. "Classe des Annélides Oligochètes" *In* "Traité de Zoologie" pp. 225–470, Vol. 5 (P. P. Grassé, ed.) Masson et C., Paris.

Cuénot, L. 1898. Etudes physiologiques sur les Oligochétes. *Arch. Biol.* 15, 79–124.

Liebmann, E. 1942. The coelomocytes of Lumbricidae. *J. Morph.*, 71, 221–249.

Metchnikoff, E. 1892. Lecons Sur la Pathologie Comparée de L'inflammation. G. Masson, Paris.

Poinar, Jr. G. O. and Leutenegger, R. 1971. Ultrastructural investigation of the melanization process in *Culex pipiens* (Culicidae) in response to a nematode. *J. Ultrastruct. Res.*, 36, 149–158.

Poinar, Jr. G. O. and Thomas, G. 1975. *Rhabditis pellio* Schneider (Nematoda) from the earthworm, *Aporrectodea trapezoides* Duges (Annelida). *J. Nematol.* 7, 374–379.

Tu, T.-J. 1937. Über die Bällchen in der Leibeshöhle der Regenwürmer. *Zool. Jb. (Anat).*, 63, 73–124.

Induction and Effector Mechanisms of Insect Immunity[1]

JUNE STEPHENS CHADWICK

DEPARTMENT OF MICROBIOLOGY AND IMMUNOLOGY
QUEEN'S UNIVERSITY
KINGSTON, ONTARIO
CANADA

---

[1]Original studies included were aided in part by Grant No. A-1958 from the National Research Council of Canada

I. INTRODUCTION

The question, "Why is insect immunity interesting" was posed
by Boman *et al.* (1974a) and, as might be expected, they brought
forth a number of satisfactory answers to the question. If we
go one step further and ask the question: "What is the basic
mechanism (or mechanisms) underlying insect immunity", we cannot
do nearly as satisfactory a job of answering.

It is well known that insects, as well as most invertebrate
species, are reasonably resistant to infection by microorganisms.
It has been shown experimentally that insects can overcome many
human pathogens which cannot be a normal part of their environment. This suggests they must possess an adequate defense
mechanism to overcome such parasites. As an example, most insect
species are not susceptible to overt infection by *Salmonella
typhi*, even though they could have little normal association with
this microorganism. Burnet (1968) has stated that, even though
insects are said to show no evidence of antibody production or
other adaptive immune characteristic, some ability to recognize
foreignness must go back a long way in evolution and the primitive building blocks for the immunoglobulins of vertebrates may
exist in invertebrates.

We apparently are still far from the answer as to what these
building blocks may be but a number of investigators are currently
producing considerable data on the "immune state" in invertebrates. The purpose of this presentation is to characterize the
state of knowledge concerning the immune response in insects,
discussing the present trends in research with specific reference
to this author's work on the induction of an effective immune
response, namely, protection against a pathogen demonstrable both
in vivo and *in vitro*. In other words, we are probably considering
what amounts to "active immunity" in the insect.

There have been several recent comprehensive reviews on insect immunity particularly those of Whitcomb *et al.* (1974) and
Cooper (1974). Therefore the situation in insects will be
summarized only to give necessary background for current trends
in research.

II. BACKGROUND INFORMATION

It is quite generally believed that the most primitive manifestation of a resistance mechanism is phagocytosis. Indeed
phagocytosis by insect hemocytes has been recognized at least
since the turn of the century, so in this respect a recognition
of foreignness clearly exists. A number of investigators,
particularly Salt (1970), have elaborated on the cellular
defense reactions of insects and there are active workers in this
field especially those interested in defense reactions to
parasitoids.

There is however, an aspect of the resistance of many insects to microorganisms which is not explicable by phagocytosis. As previously mentioned, only relatively few bacterial species are pathogenic to insects and of the many nonpathogens, it is rare to observe active phagocytosis of bacteria in the test insects on which I work, namely *Galleria mellonella*. When large numbers of bacteria are injected into the insect haemocoel, these bacteria do not rapidly disappear and can be recovered as viable organisms from the tissue or hemolymph for a considerable period of time in much the same numbers as upon introduction (Chadwick, 1973, 1975a). As we know, pathogenesis, or the lack of such, depends not only on the defense mechanisms of the host, including phagocytosis, but of course on a number of other factors, including the virulence and aggressiveness of the microorganism. It is interesting however, that in many instances the insect host makes little effort to dispose of these introduced invaders.

It is reported that the first evidence of a specific immunologic system occurs in animal species such as the hagfish (Bellanti, 1971). In these animals there occurs a high molecular weight antibody and cells responsible for a type of cell-mediated immunity. The search for a precursor to this antibody molecule goes on. Insects may not serve as a good model for such a search as they sit on the "wrong" side of the evolutionary tree with respect to vertebrates. Their innate resistance to so many microorganisms and the need to understand this resistance in terms of microbial control of noxious pests, makes the quest of understanding their immune mechanism all the more vital.

Early investigators believed insects must produce something akin to antibody because by 1920 it was recognized that an insect could be made immune to a pathogenic bacterium by vaccinating it. Within 24 hr it was possible to show that the insect could resist a hitherto lethal dose of the organism. Insect immunity according to many current investigators has the general characteristics listed in Table 1.

Many researchers automatically assumed that antibody must be produced and exhaustive serological tests were made (Chadwick, 1967). Positive claims by early workers are generally refuted today. Their results are probably explicable on grounds quite distinct from the induction of antibody.

By the 1960s it was generally believed by a number of investigators that insect immunity was a thing apart from mammalian immunity (Chadwick, 1967, 1968; Good and Papermaster, 1964). Therefore, interpretations should have henceforth been made only on the merit of "what occurs in the insect" and not on the basis of interpreting insect phenomena in mammalian terms. Unfortunately, it is easy to be victimized by the latter concept.

Table 1. General Characteristics of Induced Immunity in Insects

---

       Rapidly acquired
       Single stimulus adequate
       Brief duration
       Relatively non-specific
       Not associated with immunoglobulin
       Cell-free hemolymph shows antibacterial activity

Table 2. LD_{50s} of *Bacillus cereus* in Control and Lysozyme-
treated Groups of *Galleria mellonella* Larvae

LD_{50s} in insects treated as follows	
Lysozyme (30 min prior to vaccine)	Control
1.45×10^4	7.0×10^4
1.60×10^4	1.14×10^4
3.20×10^4	1.50×10^4

III. EXPERIMENTAL

A. *INDUCTION OF RESPONSE*

Boman *et al.* (1972) stated that in *Drosophila melanogaster* the inducer could only be a vaccine comprised of a high number of nonvirulent living bacteria and that killed organisms were ineffective. Of course these authors were describing a system of antibacterial activity against *Escherichia coli*, a nonpathogen.

In *G. mellonella* we found a rather different picture as to the nature of inducing material. Homologous heat-killed or formalized vaccines, or bacterial lipopolysaccharides, act as the most efficient immunizing agents against *Pseudomonas aeruginosa*, a pathogen (Chadwick and Vilk, 1969). However, a state of immunity can be induced by injected doses of approximately 1×10^5 *E. coli* or *Ent. aerogenes* per insect. With respect to vaccine, I had stated earlier (Stephens, 1959) that the concentration was not of great importance. Within the 10-fold concentration range we investigated this was probably true, but we have now shown that there is a minimal threshold concentration of vaccine for induction of response and a maximal concentration above which toxic effects are demonstrable in the insect (Fig. 1). When the vaccine concentration is in the order of 4×10^9 organisms/ml the insects become sluggish, darken, and may die. It is apparent also from Figure 1 that there is a minimal stimulating dose. LD_{50}s below 100 do not indicate significant immunity.

Our attempts to stimulate a response using mitogenic agents, for example, lipopolysaccharides (B cell mitogen) and Conconavalin A and Phytohemagglutinin M (T cell mitogens) resulted in inconclusive findings. Variability seems to occur depending upon the organism for which protection is being assessed and results are not reproducible. We are still in the process of investigation and interpretation of such results.

B. *SPECIFICITY*

The question of specificity in induced immunity in insects remains unsatisfactorily answered. Authors such as Wagner (1961) have felt the response was nonspecific, whereas the work of Briggs (1958), Stephens (1959), Chadwick and Vilk (1969) indicated less specificity than in higher animals, though overall some tendency to specificity was apparent. We have been interested in specificity in all facets of our investigations and, in assessing all LD_{50} results with four pathogens of *G. mellonella*, an interesting and variable pattern has evolved.

Figures 2, 3, 4, and 5 diagramatically represent a summary of our results on the efficiency of heat-killed vaccines as immunizing agents. As reported by Chadwick and Vilk (1969), good protection against *Ps. aeruginosa* is elicited by a variety of

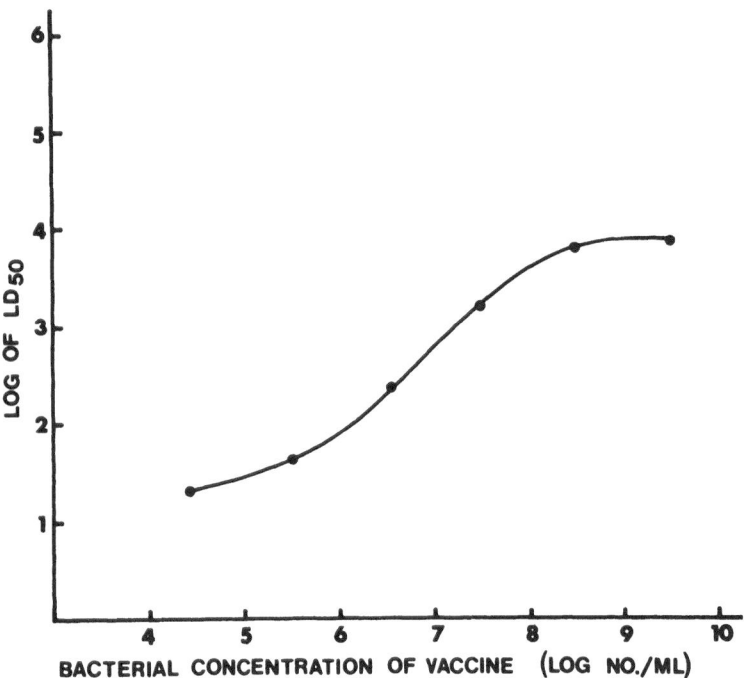

Fig. 1. Induction of immunity to *Pseudomonas aeruginosa*
Galleria mellonella by various concentrations of heat-
killed vaccines. Immunity is achieved only when LD_{50}
is at least 10 times that of control, i.e., herein
LD_{50s} above 80 indicate immunity.

vaccines and endotoxins regardless of the source (Fig. 2).   In
the case of *Proteus mirabilis* (Fig. 3) and *Serratia marcescens*
(Fig. 5), a decline in specificity accompanies the heterogeneity
of the strain or species from which the vaccine was produced.   For
Pr. rettgeri (Fig. 4) the specificity extends to homologous
species only; though not shown here, other strains of *Pr. rettgeri*
are also efficient.

C.   ANTIBACTERIAL ACTIVITY, BACTERICIDAL OR NOT?

In the known absence of antibody but with incomplete knowledge
of the factor(s) responsible for induced protection, it has, however, become apparent that the most commonly reported substance
was bactericidal (or lytic) and efforts have been made to characterize such material.

The bactericidal factor was reported to be nonprotein, of low
molecular weight, and dialysable (Stephens and Marshall, 1962).
Hink and Briggs (1968) described activity of a dual nature, one
factor apparently nonprotein and increasing on immunization, the
second low molecular weight material occurring only upon immunization. Mohrig and Messner (1968) believed all antibacterial
activity was due to lysozyme and more recently Boman *et al.* (1974b)
reported an inducible cell-free antibacterial reaction in *Samia
cecropia* pupae which while not completely described has overlapping
features with the above.

While most investigators, myself included, describe the
factor(s) as bactericidal the question arises as to whether it
truly is bactericidal or actually only inhibitory or bacteriostatic.
In instances such as the reported by Boman *et al.* (1974b) where
lysis is demonstrable, it appears that the factor involved therein
is bactericidal. In many cases the nature of antibacterial
activity has not been satisfactorily demonstrated.

In an ongoing investigation of antibacterial activity we
(Chadwick, Henderson, and Romak) have found that the activity of
"immune" wax moth hemolymph against pathogens is not immediate
but requires an induction period of 15 to 30 min. The activity is
also dependent on temperature and relative concentrations of
bacteria to hemolymph.

The above characteristics do not preclude bactericidal activity.
However, when washing or dilution techniques were used to remove
all traces of hemolymph from the treated bacteria and subsequent
daily counts were made of a suspension of washed organisms in a
pour plate, the results suggested the activity was merely
bacteriostatic. M. J. Henderson (unpubl.) found that in the case
of *Serratia marcescens* when counts of washed and unwashed bacterial
cells (previously exposed to hemolymph) were compared, the counts
of washed bacteria were higher at 24 hr and an increase again
occurred after a further 24 hr incubation. This suggested that
the depletion in numbers of bacteria occurring in the presence of

Figs. 2 & 3.   Specificity of immunity induced in *Galleria*
               mellonella against four pathogens: *Pseudomanas*
               aeruginosa 1 - 1A (Fig. 2), and *Proteus mirabilis*
               11 - 1A (Fig. 3).   Inducing agents were heat-killed
               vaccines of various bacterial species and strains.
               Length of bar is directly proportional to the mean
               Immunizing Efficiency of the agent in replicate
               experiments.

               $$\text{(Immunizing Efficiency} = \frac{\text{Experimental LD}_{50}}{\text{Control LD}_{50}})$$

               No immunity indicates an immunizing efficiency less
               than 10.

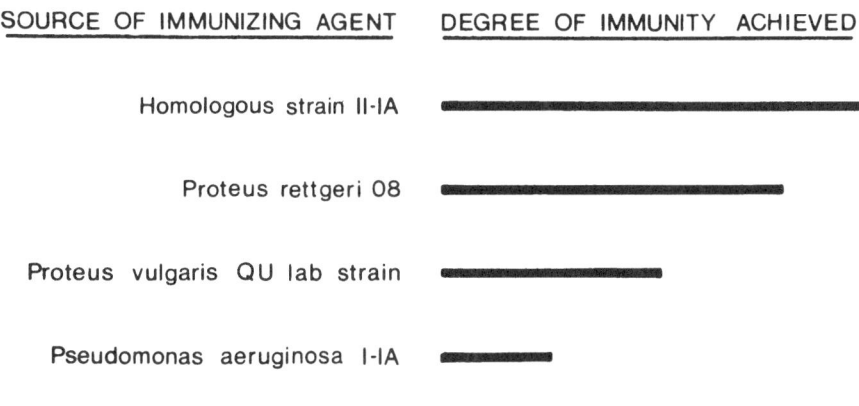

SOURCE OF IMMUNIZING AGENT   DEGREE OF IMMUNITY ACHIEVED

Homologous strain I-IA

Heterologous strain PII-I

Serratia marcescens 299

Proteus mirabilis II-IA

Proteus vulgaris QU lab strain

Homologous organism rough mutant strain   None

Figure 2

SOURCE OF IMMUNIZING AGENT   DEGREE OF IMMUNITY ACHIEVED

Homologous strain II-IA

Proteus rettgeri 08

Proteus vulgaris QU lab strain

Pseudomonas aeruginosa I-IA

Figure 3

Figs. 4 & 5. Specificity of immunity induced in *Galleria*
              mellonella against four pathogens: *Proteus*
              rettgeri 08 (Fig. 4), and *Serratia marcescens* 299
              (Fig. 5). Inducing agents were heat-killed vaccines
              of various bacterial species and strains. Length of
              bar is directly proportional to the mean Immunizing
              Efficiency of the agent in replicate experiments.

              (Immunizing Efficiency = $\dfrac{\text{Experimental } LD_{50}}{\text{Control } LD_{50}}$)

              No immunity indicates an immunizing efficiency less
              than 10.

SOURCE OF IMMUNIZING AGENT   DEGREE OF IMMUNITY ACHIEVED

        Homologous strain 08   ▬▬▬▬▬▬▬▬▬▬▬▬▬▬▬

       Proteus mirabilis II-IA   None

  Proteus vulgaris QU lab strain   None

Figure 4

SOURCE OF IMMUNIZING AGENT   DEGREE OF IMMUNITY ACHIEVED

      Homologous strain 299   ▬▬▬▬▬▬▬▬▬▬▬▬▬▬▬

Nonpathogenic S. marcescens SP-I   ▬▬▬▬

   Pseudomonas aeruginosa I-IA   ▬▬▬▬

Figure 5

hemolymph but in the absence of washing was not bactericidal but
due to the continued presence of an inhibitory factor in hemolymph.
The numbers of washed bacteria increase after 24 hr incubation
but an even greater recovery of the bacteria probably occurs in
24 - 48 hr.

Chadwick and Romak (unpubl.) found less apparent increases
when the same technique was employed with *Ps. aeruginosa*. This
may not be so surprising since this organism is probably affected
by different environmental factors than is *S. marcescens*. Also,
the immune response to *Ps. aeruginosa* is somewhat different from
that to *S. marcescens*. (Compare Figs. 2 and 5). Whether this
suggests bactericidal activity with respect to *Ps. aeruginosa*
and bacteriostatic activity against *S. marcescens* is as yet unknown. The investigation is being extended to other organisms
and under a variety of conditions in an effort to understand the
antibacterial action of immune hemolymph. The results will be
reported in detail elsewhere.

D. *LYSOZYME INVOLVEMENT AND LYTIC ACTIVITY*

Mohrig and Messner (1968) attributed all immune activity to
lysozyme whereas Powning and Davidson (1973) suggested that insect
lysozyme may be responsible for the defense against Gram-positive
organisms. Others, such as Boman *et al.* (1974b), have suggested
that lytic activity due to lysozyme may not play a large role
particularly against Gram-negative organisms but may have a more
pronounced effect on the more susceptible Gram-positive species
(Whitcomb *et al.*, 1974). Earlier, I (Chadwick, 1970) found that
while injection with either specific vaccine or nonspecific
agents would raise the lysozyme level in *G. mellonella*, this
increase in lysozyme extended well beyond the time of protective
immunity casting some doubt on the single involvement of lysozyme
as an immune factor.

In spite of technical problems due to the high LD_{50}s of most
Gram-positive bacteria for *G. mellonella*, we recently attempted
to investigate the premise of Whitcomb *et al.* (1974). Assuming
lysozyme was responsible for the destruction of Gram-positive
bacteria *in vivo*, then a prior injection of lysozyme if administered at an appropriate time might have a significant effect in
reducing the LD_{50} of Gram-positive pathogens.

Table 2 shows such was not the case. There was no significant
alteration in LD_{50}s of *B. cereus* for wax moth after lysozyme
administration. As might be expected, there was no significant
increase in protection against *Ps. aeruginosa* even though *G.
mellonella* larvae were injected with large doses of lysozyme
prior to challenge.

The answer as to the role of lysozyme in induced resistance
remains very unclear. It may well be that our timing and concentration of lysozyme dosage were not suitable to demonstrate

an effect. As to the effect of lysozyme on Gram-negative bacteria, it might be argued that immunization stimulates the release of a substance with the functional activity of complement which ultimately makes Gram-negative bacteria susceptible to lysis. This is an appealing theory and it might not be so difficult to investigate this aspect.

However, Henderson did investigate whether or not immune wax moth hemolymph had a lytic effect against *Serratia marcescens*. She was unable to show lytic activity as did Boman *et al.* (1974). Despite the fact that the hemolymph had pronounced antibacterial activity against *S. marcescens*, Henderson (unpubl.) found no change in optical density within 5 min. after addition of hemolymph to bacteria whereas Boman *et al.* (1974b) demonstrated significant decreases in optical density within 2 min after *S. cecropia* hemolymph was added to *E. coli*. The involvement of lysis may well vary from insect to insect and possibly from organism to organism.

E. SUPPRESSION OF RESPONSE BY VARIOUS AGENTS

Rasmuson *et al.* (1973) described the induction of an antibacterial factor in *Drosophila melanogaster* following vaccination with a nonpathogen such as *E. coli*. They stated that the activity, or formation, of the factor was suppressed when RNA synthesis in the insect was blocked by prior injection with Actinomycin D. Similarly, activity was suppressed by cycloheximide which presumably blocked protein synthesis. We investigated the *in vivo* effects of several well-known mammalian immunosuppressants on the induction of the response in *G. mellonella*. In each case we used as high a concentration of the agent as was soluble and that the insect would tolerate without toxic effects. The results are shown in Table 3.

Among this group of agents, the most effective in suppression of response appeared to be zymosan. Zymosan is known to have an effect on the complement pathway in vertebrates and previously I had found it interfered with *in vitro* bactericidal activity of wax moth hemolymph (Stephens, 1962).

It is interesting that most of the agents displayed some effect. While we can not draw direct analogies to their function in vertebrate systems, further research and a more exhaustive study of dosage and timing may yield useful information.

Further to our use of various immunosuppressants we investigated the effect of Cobra venom factor (CVF) on the *in vivo* and *in vitro* immune response to *Ps. aeruginosa* in *G. mellonella* (Aston *et al.*, 1976).   Cobra venom factor is known to interfere with the alternate pathway of complement in vertebrates and investigations by Day *et al.* (1970) suggested the existence of complementlike terminal components in the horseshoe crab.

We found that the administration of 0.2 or 0.4 units of CVF per insect 5 to 6 hr after vaccination and 18 hr prior to

Table 3. Suppression of Immune Response by Various Agents Administered in Conjunction with Homologous Vaccine of *Pseudomanas aeruginosa*.

Agent	LD$_{50}$s in groups injected prior to challenge with				Mean % Reduction of LD$_{50}$ by Agent
	Vaccine	Vaccine + Agent	Agent Alone	Nothing	
Actidione (Cycloheximide)	2213 1950	606.6 573.3	21.4 9.7	14.7 15.2	71.6
Chloramphenicol	5000 971	956 292	32.0 12.06	51.0 22.2	75.4
6-Mercaptopurine	2253 2370	689.6 902.2	25.3 32.2	41.2 51.1	65.7
Mitomycin C	80,009 5713	3126 1946.4	18.9 24.8	2.53 37.13	67.5
Trenomine	1127 2253	108.0 68.9	223.8 25.3	32.4 25.3	93.7
Zymosan	1074 37,210	4.26 1.00	1.0 4.3	8.8 3.8	99.8

challenge significantly lowered the LD_{50} of *Ps. aeruginosa* as compared to the LD_{50} in controls receiving vaccine only. This indicated a possible inhibitory effect by CVF on the immune mechanism. Similarly, the antibacterial capacity of immune hemolymph to *Ps. aeruginosa* was significantly decreased by incorporation of CVF in the hemolymph previous to addition of bacteria. A considerable amount of work would need to be done to clarify the role of CVF in the inhibition.

F. *CURRENT TRENDS IN RESEARCH AND IMPLICATIONS FOR THE FUTURE*

This presentation summarizes the present knowledge of some aspects of insect immunity with which the author is presently connected. There are other areas of investigation and theories which now should be mentioned not only because of their interest but also because they may overlap and contribute to the material described herein.

Altered hemolymph components and an increase in total protein as a consequence of immunity have been reported. The situation was reviewed by Chadwick (1975b). However, few investigators have thoroughly and exhaustively used separation techniques to characterize antibacterial components in hemolymph. The problem is greatly compounded because of incomplete data on normal hemolymph components. The latter may be related to wide variability which occurs not only from insect species to species but also in varying physiological states of the test insects. Certainly concentrated investigations in this area are necessary.

Moreover, research is warranted to determine how much cellular cooperation occurs in conjunction with the described humoral activity. The possible role of hemocytes in initiation of antibacterial activity remains unclear. It must be borne in mind that in some *in vivo* situations extremely virulent organisms are disposed of since the immune insect survives massive doses and this occurs in the apparent absence of phagocytosis, lysis, or a true *in vitro* bacterial effect. What then is the actual damage to the bacterial cell? To my knowledge there are few reports of this nature. Why? An analysis of the ultrastructure of such affected bacteria might provide some answers.

Table 3 reports a variety of agents which suppress the response in *G. mellonella*. Rasmuson *et al*. (1974) suggested a similar occurrence in other insect species although under somewhat different immunizing conditions. It would be interesting to determine the range of such activity and ultimately, the function of these agents which suppress an insect immune response.

A number of investigators, including Pye (1974), have drawn attention to the possible involvement of the phenoloxidase system in insect immunity. The underlying active substances are supposedly quinones which are reputedly bactericidal. According to Pye, the quinones accumulate after infection and defend the insect. Pye

also suggested that phenoloxidase activation was associated with
lysozymelike subunits. His theory, with some experimental
evidence, involves lysozyme in the immune process. Pye suggests
that lysozyme modifies foreign polysaccharide so as to expose a
site at which subunit aggregation could occur. There remains a
need for direct proof of the theory that phenoloxidase may be
involved in insect immunity though there seem few observations
to contraindicate its applicability. It may be that this theory
will be difficult to reconcile with the degree of immune
specificity observed against certain bacteria.

It is interesting too that Uhlenbruck (1974) suggested that
all invertebrate immune mechanisms recognize carbohydrate
structures. Immunity to *Ps. aeruginosa* is not induced in wax
moth larvae if the vaccine is prepared from a rough mutant
strain where polysaccharide is lacking (Fig. 2). Perhaps this
finding too is suggestive that carbohydrate is the essential
material for the induction of an immune response.

In conclusion it must be re-emphasized that variability in
response may occur from insect to insect and from pathogen to
pathogen. Therefore, even though some researcher may, in the
foreseeable future, describe an immune mechanism in chemical
terms, many other avenues remain to be explored.

REFERENCES

Aston, W. P., J. S. Chadwick, and M. J. Henderson (1976). Effect of
    Cobra Venom Factor on the *in vivo* immune response in
    Galleria mellonella to *Pseudomonas aeruginosa*. *J. Invertebr.
    Pathol.*, 27, 171-176.
Bellanti, J. A. (1971). Immunology. 585 pp., W. B. Saunders Co.,
    Philadelphia.
Boman, H. G., I. Nilsson, and B. Rasmuson (1972). Inducible
    antibacterial defence system in Drosophila. *Nature*, 237,
    232-235.
Boman, H. G., I. Nilsson-Faye, and T. Rasmuson. (1974a). Why is
    insect immunity interesting? *In* Lipmann Symposium: Energy,
    Biosynthesis and Regulation (D. Richter, ed.). Walter de
    Gruyter Verlag, Berlin - New York.
Boman, H. G., I. Nilsson-Faye, and T. Rasmuson (1974b). Insect
    immunity. 1. Characteristics of an inducible cell-free anti-
    bacterial reaction in hemolymph of *Samia cecropia* pupae.
    Infect. Immun., 10, 136-145.
Briggs, J. D. (1958). Humoral immunity in lepidopterous larvae.
    J. Exptl. Zool., 138, 155-188.
Burnet, F. M. (1968). Evolution of the immune process in verte-
    brates. *Nature*, 218, 426-430.
Chadwick, J. S. (1967). Serological responses of insects. *In*
    Symposium on Defence Reactions in Invertebrates. *Fed. Proc.
    Fed. Amer. Soc. Exp. Biol.*, 26, 1675-1679.

Chadwick, J. S. (1968). Some aspects of immune responses in
    insects. *In Vitro, 3,* 120-128.
Chadwick, J. S. (1970). Relation of lysozyme concentration to
    acquired immunity against *Pseudomonas aeruginosa* in
    Galleria mellonella. J. Invertebr. Pathol., 15, 455-456.
Chadwick, J. S. (1973). *In Vitro* growth of bacteria in tissue
    of *Galleria mellonella. J. Invertebr. Pathol., 22,* 238-241.
Chadwick, J. S. (1975a). *In Vitro* growth of bacteria in
    hemolymph of *Galleria mellonella. J. Invertebr. Pathol.,
    25,* 331-335.
Chadwick, J. S. (1975b). Hemolymph changes with infection in
    induced immunity in insects and ticks. *In* Invertebrate
    immunity (K. Maramorosch and, R. E. Shope, ed.) pp. 241-271.
    Academic Press, New York, San Francisco, London.
Chadwick, J. S. and E. Vilk (1969). Endotoxins from several
    bacterial species as immunizing agents against *Pseudomonas
    aeruginosa* in *Galleria mellonella. J. Invertebr. Pathol.,
    13,* 410-415.
Cooper, E. L. (1974). Current topics in immunobiology, Vol. 4,
    Invertebrate Immunology. 299 pp., Plenum Press, New York.
Day, N.K.B., H. Gewurz, R. Johannsen, J. Finstad, and R. A. Good
    (1970). Complement and complement-like activity in lower
    vertebrates and invertebrates. *J. Exp. Med., 132,* 941-950.
Good, R. A. and B. W. Papermaster (1964). Ontogeny and phylogeny
    of adaptive immunity. *Adv. Immunol., 4,* 1-115.
Hink, W. F. and J. D. Briggs (1968). Bactericidal factors in
    haemolymph from normal and immune wax moth larvae *Galleria
    mellonella. J. Insect Physiol., 14,* 1025-1034.
Mohrig, V. W. and B. Messner (1968). Immunreaktionen bei Insekten
    1 Lysozyme als grundlegender antibakterieller Faktor im
    humoralen Abwehrmechanismus bei Insekten. *Sond. Biol. Zent.,
    87,* 439-470.
Powning, R. F. and W. J. Davidson (1973). Studies on insect
    bacteriolytic enzymes 1· Lysozyme in haemolymph of *Galleria
    mellonella* and *Bombyx mori. Comp. Biochem. Physiol., 45B,* 669.
Pye, A. E. (1974). Microbial activation of polyphenoloxidase
    from immune insect larvae. *Nature, 251,* 610-613.
Rasmuson, B., H. G. Boman and A. Gleerup. (1973). *In vivo* in-
    duction of antibacterial defence factors in insects. Abstracts
    5th. Intern. Collog. Insect Pathol. and Microbial Control,
    Oxford p. 98.
Salt, G. (1970). The cellular defence reactions of insects.
    118 pp. Cambridge Univ. Press, Cambridge.
Stephens, J. M. (1959). Immune responses of some insects to
    some bacterial antigens. *Can. J. Microbiol., 5,* 203-228.
Stephens, J. M. and J. H. Marshall (1962). Some properties of an
    immune factor isolated from the blood of actively immunized
    wax moth larvae. *Can. J. Microbiol., 8,* 719-725.

Uhlenbruck, G. (1974).   Invertebrate immunology *In* Progr.
    Immunology 11 (2) (L. Brent and I. Holbonw, eds.)., pp.
    292-296.   North Holland Publ. Co. Amsterdam.
Wagner, R. R. (1961).   Acquired resistance to bacterial infection
    in insects. *Bacteriol. Rev.*, 25, 100-110.
Whitcomb, R. F., M. Shapiro and R. R. Granados (1974).   Insect
    defense mechanisms against microorganisms and parasitoids.
    In Physiology of Insecta (M. Rochstein, ed.), Vol. 5, p.
    447-536. Academic Press, New York.

Insect Host Responses Against Parasitoids and the Parasitoid's Resistance: With Emphasis on the Lepidoptera-Hymenoptera Association

S. Bradleigh Vinson

DEPARTMENT OF ENTOMOLOGY
TEXAS A&M UNIVERSITY
COLLEGE STATION, TEXAS

I. INTRODUCTION

The defense reactions of insect hosts towards parasitoids and the means by which parasitoids sometimes evade the host's defense have been the subject of numerous reviews (Salt, 1963a, 1968, 1970; Nappi, 1975a; Whitcomb *et al.*, 1974; Jackson *et al.*, 1969; Lafferty and Crichton, 1973). There is, however, very little consensus of opinion concerning the mechanism of the host's defense reaction or the means by which a parasitoid escapes it. As has been pointed out (Nappi, 1975; Whitcomb *et al.*, 1974; Salt, 1970), the major factors responsible for the immune response of vertebrates, notably the antigen-antibody complementary system and antigenic memory, appear to be lacking in insects.

Although the vertebrate system does not appear to operate in insects, arthropods do respond to foreign material introduced into their hemocoel. In general, insects respond to foreign materials with hemocytes either through phagocytosis of small objects or through encapsulation of objects too large to be individually engulfed. Salt (1963a) pointed out that insects usually respond to metazoan parasitoids through the encapsulation process whereby many hemocytes attach to the parasitoid surface, flatten out, and form layers which often become partly melanotic. These capsules result in the death of the parasitoid through starvation or anoxia (Fisher, 1963, 1971).

While hemocytic encapsulation is the most common means of insect defense against metazoan parasitoids, the mechanism may not be the same in all insects. Further, it would be surprising if the strategies used by different parasitoid groups with regard to the evasion of the host's defense were identical. In fact, many of the dipterous parasitoids use the encapsulation process to their advantage, diverting the process to form the respiratory sheath (Salt, 1968).

It is my purpose to examine the defense reactions of Lepidoptera using *Heliothis* spp. (Noctuidae) as an example, and to examine the evasion strategies of several hymenopterous parasitoids. This paper is not intended as a comprehensive review but will reexamine some of the results of earlier studies in view of new developments and to postulate some new hypotheses concerning the evasion of host defenses.

A. *HOST IMMUNE RESPONSE TO HYMENOPTEROUS PARASITOIDS*

While a number of *Heliothis* species serve as ovipositional sites for the braconid *Cardiochiles nigriceps* (Vinson, 1975), the parasitoid is not able to evade the host's defenses in all species (Lewis and Vinson, 1971). This process is affected by temperature and host age. Figure 1 shows the effect of temperature on the encapsulation of *C. nigriceps* eggs in 4th instar *H. zea* larvae (Lynn, 1975). The percent encapsulation at different times

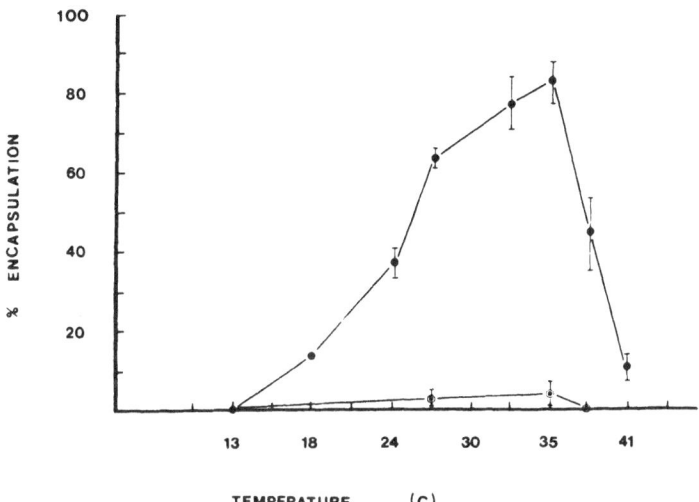

Figure 1. Effect of several temperatures on the encapsulation of *Cardiochiles nigriceps* eggs in *Heliothis zea* (represented by solid circles) 24 hr after oviposition.

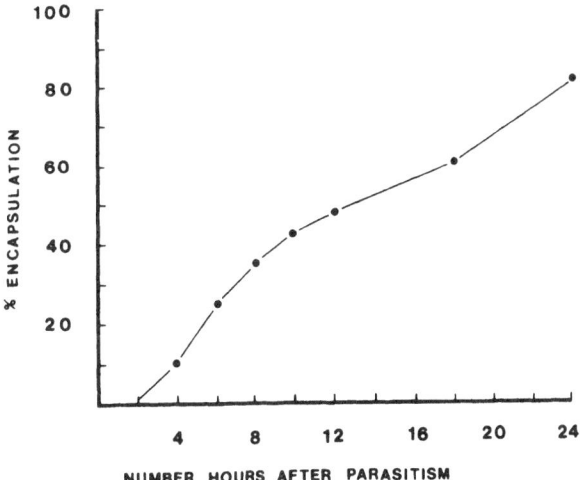

Figure 2. The percent encapsulation of *Cardiochiles nigriceps* eggs oviposited in *Heliothis zea* at different times.

after oviposition at 27°C is presented in Figure 2. Lewis and
Vinson (1968) also observed that the rate of encapsulation was
similar in both early third and late 4th instar larvae which
suggested that the molting hormones have little effect upon
encapsulation, although Nappi (1975) has suggested that a
hormonal inbalance may be a regulating factor for encapsulation.

The development of the host's defenses has been found to
increase as the host develops. Puttler (1961) found that
Peridroma saucia would encapsulate first instar larvae of the
parasitoid *Hyposoter exiguae* when oviposition occurred during
the hosts 3rd instar but the parasitoid overcame the host's
defenses if oviposition occurred during the host's 1st or 2nd
instar. It has also been found that the eggs of *C. nigriceps*
are often not encapsulated in first instar *H. zea* or *H.
philoxiphaga* (Lewis and Vinson, 1971). Walker (1959) suggested
that there was a lack of the proper type of hemocytes in young
dipterous larvae. She found that lamellocytes in *Drosophila
melanogaster* were produced earlier in the 3rd instar if
parasitized. However, the reaction of the lamellocytes depended
on the time of parasitism. If the host was parasitized during
the 1st instar, the lamellocytes produced a black pigment in the
blood which did not affect the parasitoid, while the lamallocytes
produced capsules around the parasitoid when oviposition occurred
during the second or later instars. Salt (1968) suggested that
eggs placed in the host prior to the development of the hemocytes
necessary for encapsulation, obtain a coating from the host and
are not recognized as foreign in later instars. Eggs placed
in later instars in which the phagocytic cells are developed are
attacked before they can obtain the coating. These results,
however, can be interpreted in another way. The egg or female
parasitoid could inhibit the encapsulation reaction which requires a period of time to occur. Oviposition into an early
instar host before the full development of the host's immune
response would provide the necessary time to inhibit the development of the host's defense system and the parasitoid would
escape. On the other hand, oviposition in later instar hosts
where the defense is well developed would result in the encapsulation of the parasitoid before it could inhibit the response.

There has been relatively little work with regard to a
possible response of a humoral nature involved in the hosts
defense mechanism against metazoan parasitoids. Emphasis has been
placed on the formation of the hemocyte capsule and the role
that melanins may play in the hemocyte response. Humoral
immunity has been most often described with regard to insect
responses to microorganisms (Briggs, 1958, 1964; Stephens, 1959,
1963; Gingrich, 1964; Chadwich, 1970; Hink and Briggs, 1968,
1969; Bakula, 1971; Boman *et al.*, 1972; Whitcomb *et al.*, 1974),
although relatively little is known concerning the nature of this
response.

Studies involving the response of *H. virescens* and *H. zea* and the parasitoid *C. nigriceps* have implicated a humoral involvement prior to encapsulation. Eggs of *C. nigriceps* dissected from *H. zea* prior to any evidence of hemocyte attachment and implanted in *H. virescens* were encapsulated while eggs removed from *H. virescens* and implanted in *H. virescens* were not (Lewis and Vinson, 1968). We suggested that a preparatory humoral alteration of the egg occurred in *H. zea* prior to encapsulation.

Further evidence for a humoral involvement preparatory to encapsulation was indicated by an ultrastructural examination of the eggs of *C. nigriceps* after their incubation in the two hosts (Vinson and Scott, 1974). Prior to oviposition, the egg chorion of *C. nigriceps* consists of an outer fibrous layer about 3 μm thick which is composed of small microvillar projections, and an inner chorion about 0.5 μm thick (Fig. 3a). The outer fibrous layer persist for about 6 hr after oviposition in *H. virescens* (3b) and after the loss of the microvillar projections (3c) is replaced by a more granular electron dense layer in 24 hr (Fig. 3d). Eggs in *H. virescens* hemolymph do not readily adhere to glass surfaces. Eggs deposited in *H. zea* lose the outer fibrous chorion within 2 hr along with the microvillar projections which is replaced with a microflocculate or coagulum which is condensed in 12 hr (Fig. 3ef). Eggs in *H. zea* hemolymph became "sticky" within 6 hr and readily stuck to glass surfaces, further indicating a surface change. There was no evidence of any direct hemocyte involvement in the alteration of these new layers (Vinson and Scott, 1974).

Lynn (1975) has developed further evidence for humoral involvement using *C. nigriceps*. He incubated eggs dissected from the ovary of *C. nigriceps* in the hemolymph of 4th instars of *H. zea* for 1 hr, rinsed them in distilled water, and injected them into *H. virescens*. The procedure was repeated with *H. virescens* hemolymph, and heated hemolymph of both species. The results, given in Table 1, show that 18% of the eggs incubated in *H. virescens* hemolymph *in vitro* were encapsulated in *H. virescens*, which may be the result of manipulation. In contrast, over 60% of the eggs incubated in *H. zea* hemolymph *in vitro* were encapsulated. Furthermore, the activity in the *H. zea* hemolymph was lost upon heating, indicating the protein nature of the factor involved. No change occurred under these conditions with *H. virescens* hemolymph. Since this evidence suggested that *H. zea* hemolymph had a factor lacking in *H. virescens*, Brewer *et al.* (1972) examined the possibility of this factor being a protein by using antibody absorption techniques. He demonstrated the presence of an antigen in *H. zea* that was absent in *H. virescens*.

In contrast to *C. nigriceps*, the ichneumonid *Campoletis sonorensis* attacks and survives in both *H. virescens* and

Table 1. The encapsulation of *Cardiochiles nigriceps* eggs in
          Heliothis virescens after their *in vitro* incubation in
          variously treated hemolymph.

Test Hemolymph	Number Eggs Recovered	Mean Percent Encapsulation	S.E.
H. zea	15	73.3	11.5
H. virescens	28	14.3	8.3
Heated *H. zea*	19	10.5	5.7
Heated *H. virescens*	21	19.0	9.1

Table 2. Percent encapsulation of *Cardiochiles nigriceps* eggs in
          Heliothis zea parasitized 1 hr prior by indicated species

Treatment	No. Larvae dissected	No. eggs recovered	% Enc.	± S.E.
C. nigriceps	12	24	75.0	5.29
C. sonorensis	29	29	26.6	8.24
M. crociepes	23	22	9.1	6.98
M. crociepes fluid[1]	16	16	12.5	7.08
Saline[1]	14	14	85.7	2.38

1/ Larvae injected with 1 μl calyx from *M. crociepes* or insect
   saline.

H. zea. The egg chorion is structurally similar to that of *C. nigriceps* (Fig. 4ab) but the outer fibrous layer is lost in *H. zea* while the microvillar projections remain (Fig. 4b). *C. sonorensis* eggs incubated in protease or lipase were readily encapsulated by *H. virescens* although no observable change in the chorion could be noted. After encapsulation, a thin, electron dense material could be seen between the hemocyte surface and the chorion of the egg (Fig. 4c) (Norton and Vinson, unpubl.). These observations demonstrate the different responses of a host to the eggs of different parasitoid species.

A second line of evidence suggesting a humoral involvement is the possible role of the tyrosine-phenoloxidase system in the immune process. The importance of the phenoloxidase system has been discussed by a number of authors (Taylor, 1969; Poinar *et al.*, 1968; Götz, 1973; Götz and Vey, 1975; Nappi, 1973a, 1975a). Most of the discussion has centered on the role of the melanins. The melanins are an ubiquitous group of compounds that are the result of the oxidation and polymerization of a number of phenols through the action of phenoloxidases, phenolase, and tyrosinase (Nicolaus, 1968). The phenoloxidases are important in both the darkening (melanization) and hardening (scleroti-zation) of the cuticle (Hackman, 1971). Both processes can occur independently of each other (Hackman, 1974), although both processes involve many of the same precursors.

While melanotic capsules around parasitoids have been de-scribed (Salt, 1956), melanization has often been ascribed to a secondary role after encapsulation by the hemocytes (Muldrew, 1953; Bronskill, 1960; Salt, 1963, 1965). A more active role for the tyrosine-phenoloxidase system has been proposed by several authors (Taylor 1969; Götz and Vey 1974; Nappi 1975a). In Diptera where melantic capsules form in the absence of a direct cellular reaction (Bronskill, 1962; Essinger, 1962; Götz, 1973), the importance of melanin has been well demonstrated (Götz and Vey, 1974). The importance of the phenoloxidase system in preparing the foreign surface for encapsulation by hemocytes has also been indicated. By injecting phenylthiourea (PTU) or re-duced glutathione into *H. zea,* Brewer and Vinson (1971) observed that encapsulation was reduced. Nappi (1973a) found that en-capsulation of *Pseudeucoila bochei* was reduced in *D. melanogaster* fed PTU in the diet.

Although the tryosine-phenoloxidase system may be involved in the hosts defense against parasitoids, whether this system acts to kill the parasitoid prior to encapsulation or acts more subtly by altering the parasitoid surface so that the hemocytes can attack is unknown. Nappi (1973) showed that eggs of *P. bochei* were killed prior to melanization and encapsulation. Salt (1956) found that living eggs and larvae were heavily encapsulated in *Carausius morosus.* Other examples are given by Salt (1963a). The role of the phenoloxidase system in immunity particularly

Figure 3.   (A) Egg of *Cardiochiles nigriceps* in the female calyx.
             Note the viruslike particles (P). (B) Egg of *C.
             nigriceps* in susceptible *Heliothis virescens* for 6 hr
             showing loss of most of the fibrous layer (F) and the
             presence of small dense particles (DP). (C) Egg of
             C. nigriceps in *H. virescens* for 12 hr showing complete
             loss of the fibrous layer. (*D*) Egg of C. *nigriceps*
             in *H. virescens* for 24 hr showing the thin granular
             layer (GL). (E) Egg of *C. nigriceps* in *H. zea* for 2
             hr showing the loss of a fibrous layer and the presence
             of a microfloculate (MF). (F) Egg of *C. nigriceps* in
             H. zea for 24 hr showing a condensing in the micro-
             floculate. CMF = condensed microfloculate material;
             DP = dense particles; E = endochorion; F = fibrous
             layer; GL = granular layer; IR = irregular layer;
             M = microvilli; P = calyx particles.

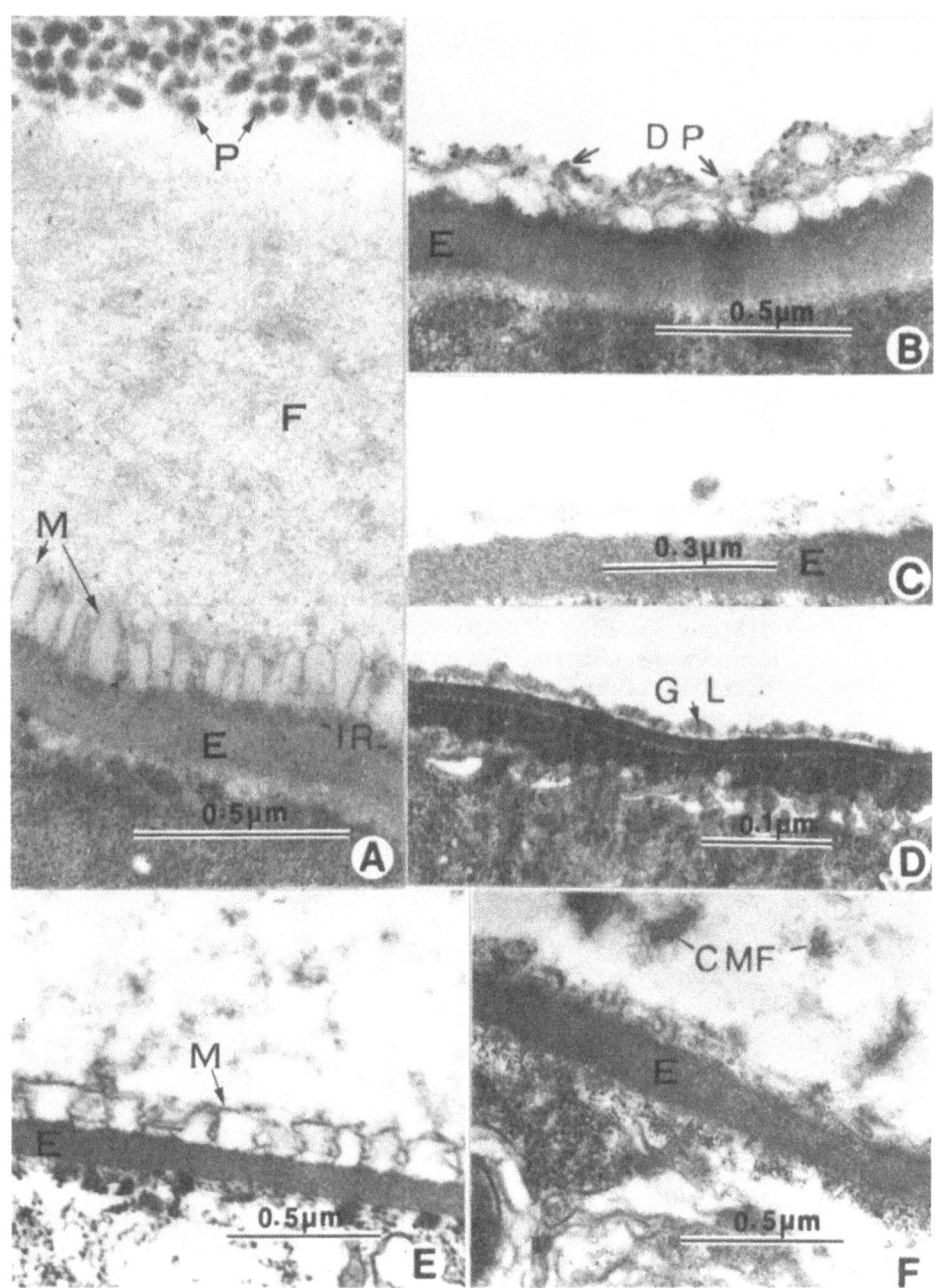

Figure 4.   (A) The egg of *Campoletis sonorensis* in the female
            lateral oviduct showing a thin fibrous layer (F) and
            the oval particles (P). (B) The egg of *C. sonorensis*
            in *H. virescens* for 1 hr showing the absence of the
            fibrous layer. (C) The encapsulated egg of *C.
            sonorensis* showing the presence of a denser granular
            material around the microvillar projections. E =
            endochorion; F = fibrous layer; H = hemocyte; M =
            microvilli and P = calyx particle.

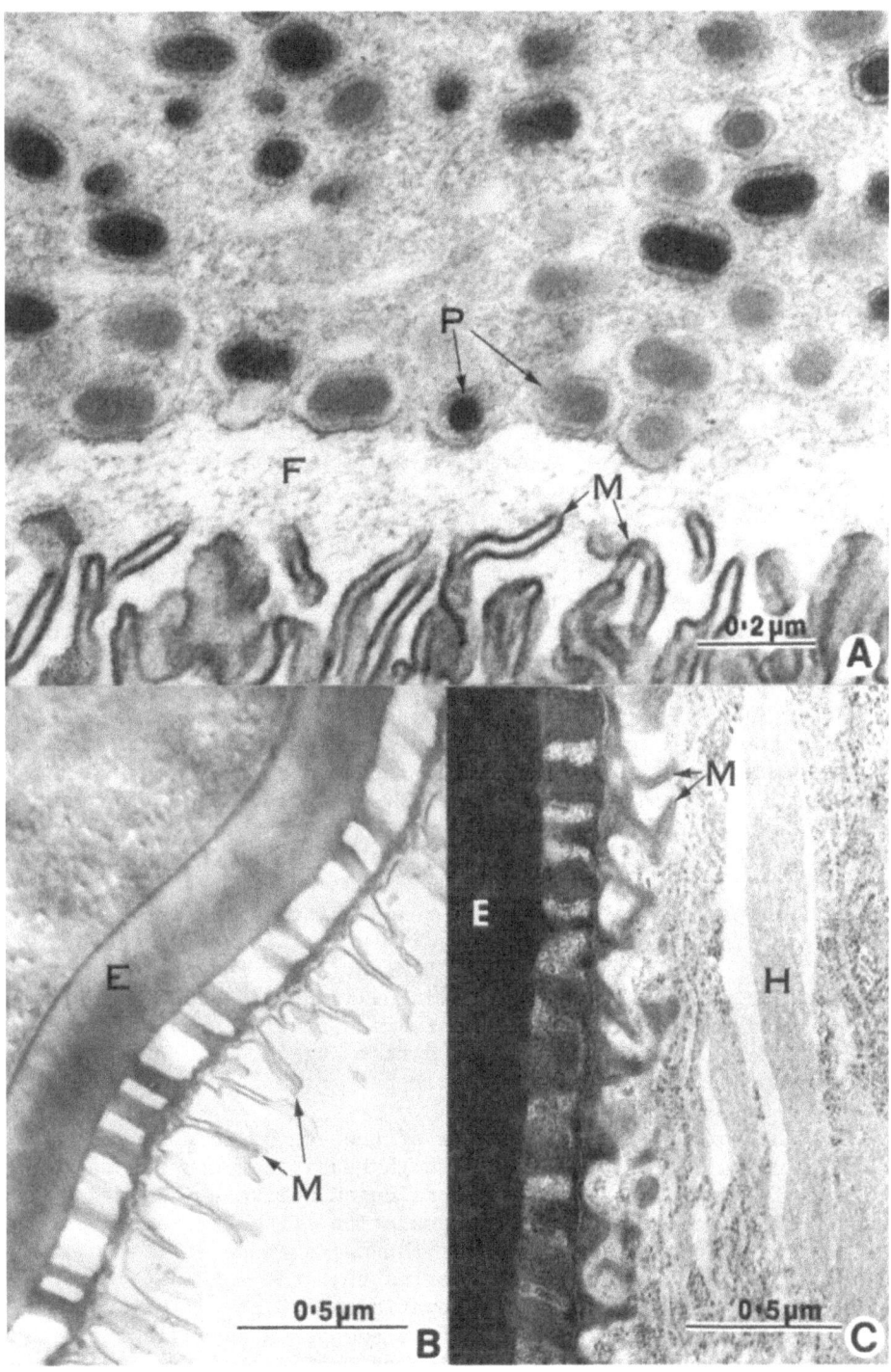

with regard to its possible role prior to hemocyte attachment
appears ill-defined.

B. *THE EVASION OF THE HOST IMMUNE RESPONSE BY HYMENOPTEROUS*
    PARASITOIDS

The evasion of host defenses by parasitoids may represent a
set of strategies as diverse as the taxonomic orders involved.
Further, coevolution of the host-parasitoid interaction may have
resulted in a multiple system. As the parasitoid overcomes the
hosts defenses, the host responds by increasing or altering its
defenses which are then countered by the parasitoid.

There are a number of possible strategies of an active or
passive nature that may be utilized by insect parasitoids which
inhabit the host's hemocoel. The passive strategies may include
the acquisition of host materials which coat the parasitoid,
resulting in the failure of the host to recognize the parasitoid
as foreign; the possession of heterophilic antigens (Rowley and
Jenkins, 1962) or use of molecular mimicry (Damian 1965); or the
possession of nonreactive surface compounds of, for example,
mucopolysaccharides. Active strategies include repulsion of
hemocytes, secretion of a toxin or agents which neutralize or
interfere with the hemocytic response, suppression of hemocyte
transformation, or sequestering of materials (attrition)
necessary to carry out a hemocytic reaction.

Salt (1965, 1966) suggested that the host fails to recognize
the parasitoid egg as foreign due to the presence of a coating,
possibly a mucopolysaccharide that is not recognized or, in some
other way, fails to elicit a response. A similar suggestion was
made by Lewis and Vinson (1968).

Salt (1965) provided evidence that the egg of *Venturia*
(=*Nemeritis*) *canescens* in *Anagasta* (=*Ephestia*) *kuehniella*
was not encapsulated while other foreign objects implanted into
the hemocoel of this species were quickly encapsulated. Similar
results were provided by Lewis and Vinson (1968) for the eggs of
C. nigriceps which were not encapsulated in *H. virescens* while
wood chips implanted into these hosts were readily encapsulated.
In both cases, evidence was provided that suggested that the
eggs did not secrete compounds that repel or inhibit the hemocytes.

Chemical or physical alteration of the egg by abrasion or
chemical means (Salt, 1965), or through heat treatment or coating
the egg with antibodies from egg challenged chickens (Lewis and
Vinson, 1968) resulted in the encapsulation of the implanted eggs.
These results show that a change in the egg surface will result
in encapsulation. Salt (1965) reported that the egg was altered
in some way as it passed from the ovariole into the calyx, since
eggs dissected from the ovarioles were encapsulated while eggs
from the calyx were not. Examination of the eggs of *V. canescens*

revealed a thick translucent or fibrous layer covering eggs from
the calyx but only thinly coating eggs taken from the ovariole
(Rotheram, 1973a). Rotheram (1967, 1973ab) also found numerous
particles in the calyx and on the egg surface which Rotheram
(1973b) and Salt (1973) have suggested protect the egg by coating
its surface.

An amorphous or fibrous layer surrounding eggs of other
parasitoids has also been reported (King and Richards, 1969;
King *et al.*, 1969; Vinson and Scott, 1974; Norton *et al.*, 1974).
More recently, Osman (1974) has provided evidence that the eggs
of *Pimpla turionellae* are coated with a mucopolysaccharide and
using rethinum red, we have observed that the outer fibrous layer
of *C. nigriceps* shows some staining and that the microvillar
layer stains intensely indicating the presence of mucopolysaccharides (Vinson and Scott, unpubl.). Fuhrer (1973) also
reported mucopolysaccharides in the accessory glands of *P.
turionellae*. As pointed out by Salt (1970), mucopolysaccharides
may not elicit a response from insect hemocytes.

It is of interest to note that mucopolysaccharides are acidic
compounds. It was found that the eggs of *C. nigriceps, M.
croceipes*, and an acidic cationic exchanger, Sephadex C-50, were
not readily encapsulated in *H. virescens* and did not react with
each other or with vital red stain. Walters and Williams (1966)
also observed that hemocytes were unable to phagocytose
cationic exchange granules. In contrast the basic anionic
exchanger, Sephadex A-50, reacted with vital red and the eggs of
the parasitoids as well as with the calyx fluid of the two
parasitoids. The A-50 Sephadex failed to react to methylene blue
stain while the C-50 did. Damaged host tissue, nylon fibers,
glass beads and wax pellets reacted with both stains and were
encapsulated (Vinson, 1974).

Surfaces with mixed ionic sites, such as nylon fibers, may
allow for the attachment of proteins or some other material,
preparing the surface for subsequent hemocyte attachment. A
surface with few positive (cationic) sites may not provide
enough sites for the attachment of an opsonin-like substance.

If mucopolysaccharides afford protection, then why are the
eggs of *C. nigriceps* encapsulated in *H. zea?* Vinson and Scott
(1974) observed that the egg of *C. nigriceps* oviposited in
H. zea lost the outer amorphous layer within 2 hr while the
fibrous outer layer of eggs in the hemocoel of *H. virescens*
was slowly lost over 12 hr. During the loss of the outer
fibrous layer of the chorion in *H. virescens*, a number of small
particles appeared embedded in the outer layers of the chorion.
These may be involved in the digestion or dissolution of the layer.
A difference in the dissolution rate of the outer fibrous layer
may contribute to the different host responses. Once the outer
layer is lost in *H. zea* a new layer is observed which may prepare
the surface for encapsulation. The slow loss of the outer

fibrous layer of *C. nigriceps* in *H. virescens* may provide the
time necessary for the calyx particles (discussed below) to
regulate the host. Host regulation is much more rapid with
C. sonorensis, possibly reducing the need for a protective
surface. Also the microvillar projections of the egg of *C.
sonorensis* appear more resistant to dissolution and remain after
their incubation in the host affording protection (Norton and
Vinson, 1975).

A number of other theories have been advanced concerning
a mechanism for the evasion of the host's defenses by insect
parasitoids which appear to be directed more toward the larval
stage of the parasitoid. Salt (1968) has suggested that depletion of the host's food reserves or hemocytes may account for
the protection of parasitoid larvae. Salt (1968, 1971) postulated that teratocytes, arising from an embryonic membrane of the
parasitoid which, upon egg-hatch, dissociates into individual
cells which rapidly increase in size, deplete the hemolymph of food
materials and thus weaken the host's defense. The concept of
attrition and weakening of the host's defense would appear to be
nonspecific. As mentioned by Vinson (1972a), parasitized hosts
continue to respond to other foreign objects which invade the
hemocoel. For example, Vinson and Lewis (1973) found that the
host's which remain alive after *Microplitis crociepes* larvae
emerge, will encapsulate brush chips, although the hemocoel
contains many teratocytes. Also, teratocytes injected into *H. zea*
were encapsulated while teratocytes injected into *H. virescens* continued to grow (Vinxon, 1970). The role of teratocytes in the
parasitoid-host association is still undetermined. Vinson (1972a)
found that young larvae of *C. nigriceps* which were encapsulated
when injected into another potential *H. virescens* host were not
if several teratocytes were first injected.

Another possible means of evading the host defense system is
through the injection of some material that suppresses part of
the hosts defense. Such a system would have to account for the
specificity which is shown, for example, by *C. nigriceps* which is
not encapsulated in *H. virescens* but is, in the closely related
H. zea (Lewis and Vinson, 1968). It must account for the fact
that the material does not turn off the complete system since
parasitized *H. virescens* larvae can still encapsulate certain
other foreign objects (Vinson, 1972a). It must also account for
the fact that *C. sonorensis* and *M. croceipes* can escape or
suppress the encapsulation reaction in both *H. virescens* and *H.
zea*, yet the parasitized hosts still encapsulate other foreign
material. While such a mechanism cannot yet be explained there
are some facts that can be used to develop a hypothesis.

Pemberton and Willard (1918) reported that larvae of the
chalcid *Tetrastichus giffordianus* were encapsulated in larvae of
Dacus cucurbitae when they occurred alone but developed in hosts
attacked by the braconid, *Opius fletcheri*. They postulated that

the braconid injected a toxic material that suppressed the host's defenses thus allowing for the development of the chalcid. More recently, Streams and Greenberg (1969) reported that *Pseudeucoila mellipes* are normally encapsulated in *D. melanogaster* but are protected if hosts are also infected with *P. bochei* which normally survives in *D. malanogaster*. These authors suggested that the female *P. bochei* injected a substance or the egg secreted something that protected *P. mellipes*. The fact that this protection failed if the infection by *P. bochei* was delayed over 36 hr indicates that the factor requires a period of time to become active.

Larvae of *C. nigriceps* removed from a host and implanted into *H. virescens* are normally encapsulated, yet they are protected if the recipient *H. virescens* has been parasitized by a female *C. nigriceps* or injected with the females calyx fluid (Vinson, 1972a). Another example of this type of protection is given in Table 2. For this study, *H. zea* was parasitized by *C. nigriceps*, *C. sonorensis*, or *M. croceipes* and 1 hr later, by *C. nigriceps*. The eggs of *C. nigriceps* were readily encapsulated in *H. zea* parasitized by *C. nigriceps*, less so if parasitized by *C. sonorensis*, and protected if parasitized by *M. croceipes* (unpubl.).

Similar protection of *C. nigriceps* eggs in *H. zea* occurred if the host was first injected with calyx fluid from *M. croceipes* but not with the saline carrier alone (unpubl.). Unlike those of Rotheram (1973a) and Salt (1973), these studies, and the observation that the calyx particles of *C. nigriceps* and *C. sonorensis* do not coat the egg surface (Fig. 3a, 4a), indicate that the calyx fluid must in some way suppress the ability of hemocytes to form capsules specifically around insect parasitoids while retaining the ability to encapsulate other foreign surfaces.

Whether the calyx particles act at the surface (Salt, 1973) or on some system in the host which suppresses the production of compounds necessary for preparing the foreign surface for encapsulation or by suppressing hemocyte transformation is unknown. A number of authors have reported alteration or differential suppression of certain hemocyte types after parasitism (Walker, 1959; Nappi and Streams, 1969; Nappi, 1975b; Kitano, 1974). Nappi (1973b) found that there was an increase in the formation of lamellocytes in susceptible *D. algonquin* in response to *P. bochei*, indicating host recognition of the foreign material but the hemocytes failed to encapsulate. Vinson (1971) found no changes in differential hemocyte counts during the first few days of parasitism, thus the suppression of a specific hemocyte type did not seem to play a role in the defense of the egg. An alteration of the hemocyte composition may play a role in the defense of larvae since there was a drop in spherule cell numbers in 3 days. However, the significance of the alteration of differential hemocyte counts in the host is obscured if these hosts remain capable of encapsulating other foreign material, such as brush

chips or glass beads, (Vinson, 1972a). It would be interesting
to know if hosts in which the suppression of hemocyte transformation has been suggested as an important factor in the evasion
of host defenses are capable of the encapsulation of other foreign
material introduced into the hemocoel of parasitized hosts.

C. ROLE OF PARTICLES ASSOCIATED WITH CERTAIN PARASITOIDS

Rotheram (1973a,b) reported that the calyx of *V. canescens*
contained a heterogeneous liquid, the particles of which were about
130 μm in diameter. She observed that these particles were attached to and embedded in the outer layer of the egg. Similar
particles were also observed on the outer cuticle of first instar
larvae (Rotheram, 1973a). Salt (1973) suggested that these
particles in *V. canescens* coat the surface of the egg and larvae
and in some way, protect them from encapsulation.

Vinson (1969) reported that the calyx fluid of *C. nigriceps*
had suspended in it what appeared to be particulate matter. An
electron microscopical study of the eggs and calyx region of the
reproductive system of *C. nigriceps* revealed the presence of
numerous particles (Fig. 3a) which surrounded the egg (Vinson
and Scott, 1974). Norton *et al*. (1975) have observed similar
particles surrounding the egg of *C. sonorensis* in the lateral
oviduct (Fig. 4a). However, in contrast to the studies of
Rotheram (1973a), the particles in *C. nigriceps* and *C. sonorensis*
did not appear to be attached or embedded in the outer chorion
of the egg. Furthermore, no evidence of these particles was
observed on the eggs of either of these parasitoids when they were
removed from hosts 2 hr after oviposition (Vinson and Scott, 1975;
Norton *et al*., 1975).

These particles are produced in the nucleus of unusually
enlarged cells in the calyx or along the calyx wall (Rotheram,
1973; Vinson and Scott, 1974; Norton *et al*., 1975; Stoltz *et al*.,
1975). In *C. sonorensis* and *N. canescens* these particles migrate
from the nucleus through the cytoplasm and are secreted into the
calyx lumen (Norton *et al*., 1975; Rotheram, 1973a), while in
C. nigriceps the nucleus enlarges, almost filling the cell which
then ruptures liberating the particles (Vinson and Scott, 1975).

There appears to be considerable variation in the particles
of the species thus far investigated. Rotheram (1967, 1973b)
reported that the particles appeared angular with several lobed
profiles and were bounded by a unit membrane. The outer membrane
stained for acid mucopolysaccharides as did the egg's outer
chorion. She also obtained no evidence for the presence of DNA
nor did Bedwin (see Rotheram, 1973b).

The particles from *C. nigriceps* are about 130 μm in cross
section and appear oval. They consist of a densely staining inner
core surrounded by an amorphous material and enclosed by a unit
membrane. The central core contains DNA (Vinson and Scott, 1975)

and the outer membrane is coated with an amorphous material which stains with rethimum red, indicative of mucopolysaccharide. The particles in *C. sonorensis* consist of an inner core 0.25 μm by 0.068 μm surrounded by a membrane with a varying amount of space between the core and membrane (Norton *et al.*, 1975). Stoltz *et al.* (1975) has recently examined the particles from three species of braconids, including *C. nigriceps*, and on the basis of apparent structural homologies has suggested that the particles in the braconid species examined thus far appear related to baculoviruses.

The calyx fluid has been found to cause a number of pathologies. Jones and Lewis (1971) found that the calyx fluid of *M. croceipes* suppressed host growth and respiration. Similar results were found for *C. nigriceps* and *C. sonorensis* (Vinson, 1972b; Vinson and Barras, 1970). Injection of calyx fluid from *C. nigriceps* resulted in suppression of growth in 5 days while fluid from the lateral oviducts of *C. sonorensis* caused a measurable suppression of growth in one day (Vinson and Barras, 1970; Vinson, 1972b). Further, Dahlman and Vinson (1975) have found that parasitism of *H. virescens* by *M. croceipes* elevates the trehalose level of the host's hemolymph while *C. nigriceps* does not. The elevation of hemolymph trehalose in *H. virescens* has been found to occur following injection of the calyx fluid of *M. croceipes* (Dahlman and Vinson, unpubl). What other changes may occur is unknown but the results indicate a degree of specificity in the effects elicited by the female parasitoid.

Salt (1973) has suggested that the particles act as an immunosuppressive agent acting at the surface. He states they (1) suppress the reaction of hemocytes to a foreign body, (2) act on specific blood cell reactions not their proliferation, and (3) they do not act through a general toxic effect. In contrast, the particles in *C. nigriceps* and *C. sonorensis* affect a number of host systems. The injection of a DNA containing virus-like particles could regulate a number of physiological systems of the host. The possibility that parasitoids have developed a symbiotic relationship with a virus which is capable of suppressing the hosts immune response and altering various other physiological systems to the parasitoids advantage opens up a facinating field for investigation.

SUMMARY

The defense of Lepidoptera, primarily *Heliothis* species, against hymenopterous parasitoids and the parasitoid's strategies at counteracting the hosts defense are discussed. The evidence of a host-borne humoral response which may involve the phenoloxidase system prior to encapsulation by the hosts hemocytes is presented.

The parasitoid egg may escape the hosts defense due to the presence of a surface coating, possibly a mucopolysaccharide, which is susceptible to degrees of degradation by the host. The later egg and larval stages of the parasitoid may derive protection from particles injected by the female parasitoid during oviposition. These particles are a form of virus and appear to be related to the baculoviruses.

ACKNOWLEDGMENT

I wish to thank Dr. G. F. Iwantsch and Mr. D. C. Lynn for their valuable review and suggestions. Approved for publication as TA___12093___ by the Director, Texas Agricultural Experiment Station and conducted in cooperation with the Agricultural Research Service, USDA.

REFERENCES

Bakula, M. (1971). The isolation of intracellular antibacterial activity from *Drosophila melanogaster* larvae. *J. Invest. Physiol.*, 17, 313–319.

Brewer, F. D. and S. B. Vinson. (1971). Chemicals affecting the encapsulations of foreign material in an insect. *J. Invertebr. Pathol.*, 18, 287–289.

Brewer, F. D., B. Glick, and S. B. Vinson. (1972). Immunological investigations of the factor(s) responsible for the resistance of *Heliothis virescens* to the parasitoid *Cardiochiles nigriceps*. *Comp. Biochem. Physiol.*, 43, (4B), 781–786.

Boman, H. G., Nilsson, I., and Rasmuson, B. (1972). Inducible antibacterial defense system in drosophila. *Natur. Lond.*, 237, 232–235.

Briggs, J. D. (1958). Humoral immunity in lepidopterous larvae. *Jour. Exptl. Zool.*, 138, 155–88.

Briggs, J. D. (1964). "Immunological Responses" *In* "The Physiology of Insects," M. Rockstein (ed.). Academic Press. New York.

Bronskill, J. F. (1960). The capsule and its relation to the embryogensis of the ichneumonid parasitoid *Mesoleius tenthredinis* Morl. in the larch sawfly, *Pristiphora erichsoni* (Htg.) (Hymenoptera, Tenthredinidae) *Canad. J. Zoolo.* 38, 769–775.

Chadwick, J. S. (1970). Relation of Lysozyme concentration to acquired immunity against *Pseudomonas aeruginosa Galleria mellonella*. *J. Invertebr. Pathol.*, 15, 455–456.

Dahlman, D. L. and S. B. Vinson. (1975). Trehalose and glucose levels in the Hemolymph of *Heliothis virescens* parasitized by *Microplitis croceipes* or *Cardrochiles nigriceps*. *Comp. Biochem. Physiol.* 52B, 465–468.

Damian, R. T. (1965). Molecular mimicry: antigen sharing by parasite and host and its consequences. *Amer. Nat.*, 118, 129-250.

Essinger, J. H. (1962). Behavior of microfilariae of *Brugea pahangi* in *Anopheles quadrimaculatus*. *Amer. J. Trop. Med. Hyg.*, 11, 749-758.

Fisher, R. C. (1963). Oxygen requirements and the physiological suppression of supernumerary insect parasitoids. *J. Exp. Biol.*, 40, 531-540.

Fisher, R. C. (1971). Aspects of the physiology of endoparasitic Hymenoptera. *Biol. Rev.*, 46, 243-278.

Führer, E. (1973). Sekretion von Mucopolysacchariden im weiblichen Geschlechtsapparat von *Pimpla turionellae* L. (Hym., Ichneumonidae) Zeit. *Parasitenk.*, 41, 207-214.

Gingrich, N. E. (1964). Acquired humoral response of the large milkweed bug, *Oncopeltus* fasiatus (Dallas) to injected materials. *J. Insect Physiol.*, 10, 179-184.

Götz, P. (1973). Immunreaktionen bei Insekten. *Naturw. Rdsch.*, 26, 367-375.

Götz, P. and A. Vey. (1974). Humoral encapsulation in Diptera (Insecta): defence reactions of *Chironomus* larvae against fungi. *Parasitology*, 68, 193-205.

Hackman, R. H. (1971). The Integument of Arthropods. *In* "Chemical Zoology" (M. Florkin and B. T. Scheer, ed.) Vol 6B. pp. 1-62. Academic Press, New York.

Hackman, R. H. (1974. Chemistry of the insect cuticle *In* "The physiology of insecta" (Rockstein, M., ed) Vol. 5. pp. 215-270. Academic Press, New York.

Hink, W. F. and Briggs, J. D. (1968). Bactericidal factors in haemolymph from normal and immune wax moth larvae *Galleria mellonella*. *J. Insect Physiol.*, 14, 1025-1034.

Hink, W. F. and Briggs, J. D. (1969). Immune responses of ligatured *Galleria mellonella* larvae. *J. Insect Pathol.*, 13, 308-309.

Jackson, G. J., Herman, R. and Singer, I. (1969). "Immunity to parasitic animals" Vol. 1. Appleton Century Crofts, New York.

Jones, R. L. and W. J. Lewis. (1971). Physiology of the host-parasite relationship between *Heliothis zea* and *Microplites croceipes* *J. Insect Physiol.*, 17, 921-927.

King, P. E. and Richards, J. G. (1969). Oogenesis in *Nasonia vitripennis* (Walker) (Hymenoptera: Pteromalidae). *Proc. R. Ento. Soc. Lond. (A)*, 44, 143-157.

King, P. E., Ratcliffe, N. A. and Copland, M.J.W. (1969). The structure of the egg membranes in *Apanteles glomeratus* (L.) (Hymenoptera: Braconidae). *Proc. R. Ent. Soc. Lond. (A)*, 44, 137-142.

Kitano, H. (1974). Effects of the parasitization of a Braconid, *Apanteles*, on the blood of it's host, *Pieris*. *J. Insect Physiol.*, 20, 315-327.

Lafferty, K. J. and R. Crichton. (1973). Immune responses of
    invertebrates *In* "Viruses and invertebrates". (A. J. Gibbs,
    ed.) pp. 301-320, Frontiers of Biology Vol. 31. American
    Elsevier Publ., New York.
Lewis, W. J. and S. B. Vinson. (1968). Immunological relationships
    between the parasite *Cardiodhiles nigriceps* Viereck and
    Certain *Heliothis spp.* *J. Insect Physiol.*, 14, 613-626.
Lewis, W. J. and S. B. Vinson. (1971). Suitability of certain
    Heliothis (Lepidoptera: Noctuidae) as hosts for the parasite
    Cardiochiles nigriceps. *Ann. Entomol. Soc. Am.*, 64, 970-972.
Lynn, D. C. (1975). Biochemical differences between *Heliothis zea*
    (Boddie) and *Heliothis virescens* (F.) with respect to their
    internal defense against the parasitoid *Cardiochiles nigriceps*
    (Vierick). *Thesis Texas A & M University.*
Muldrew, J. A. (1953). The natural immunity of the larch sawfly
    (Pristiphora erichsonii) to the introduced parasite
    Mesoleius tenthredinis Can. J. Zool., 31, 313-332.
Nappi, A. J. (1973a). The role of melanization in the immune
    reaction of larvae of *Drosophila algonguin* against *Pseudeu-
    coila bochei. Parasitology*, 66, 23-32.
Nappi, A. J. (1973b). Hemocytic changes associated with the
    encapsulation and melanization of some insect parasites.
    Exp. Parasit. 33, 285-302.
Nappi, A. J. (1975a). Parasite encapsulation in insects *In*
    "Invertebrate Immunity" (Maramorosch. K and Shope, R. E.,
    ed.) Academic Press.
Nappi, A. J. (1975b). Cellular immune-reactions of larvae of
    Drosophila algonquin. Parasitology, 70, 189-194.
Nappi, A. J. and Streams, F. A. (1969). Haemocytic reactions of
    Drosophila melanogaster to the parasites *Pseudeucoila mellipes*
    and *P. bochei. J. Insect Physiol.*, 15, 1551-1566.
Nicholaus, R. A. (1968). "Melanins." Hermann, Paris.
Norton, W. N., S. B. Vinson and E. L. Thurston. (1974). An
    ultrastructural study of the female reproduction tract of
    Campolitis sonorensis (Hymenoptera). *Proc. Electron.
    Microscopy Sco.*, 1974, 140-141.
Norton, W. N., S. B. Vinton and D. B. Stoltz. (1975). Nuclear
    secretory particles associated with the calyx cells of the
    ichneumonid parasitoid *Campoletis sonorensis* (Cameron).
    Cell and Tissue Res., 162, 195-208.
Osman, S. E. (1974). Parasitentoleranz van Schmetterlingspuppen
    Maskierung der Parasiteneier mit Mucopolysacchariden.
    Naturwissenschaften, 61, 453-454.
Pemberton, C. E. and Willard, H. F. (1918). A contribution to the
    biology of fruit-fly parasites in Hawaii. *J. Agric. Res.*,
    15, 419-466.
Poinar, G. O., Jr., Lentenegger, R. and Götz, P. (1968). Ultra-
    structure of the formation of a melanotic capsule in *Diabrotica*
    (Coleoptera) in response to a parasitic nematode (Mermithidae).
    J. Ulstruct. Res. 25, 293-306.

Puttler, B. (1961). Biology of *Hyposter exiguae* (Hymenoptera:
Ichneumonidae), a parasite of lepidopterous larvae. *Ann.
Entomol. Soc. Am.*, 54, 25-30.

Rotheram, S. (1967). Immune surface of eggs of a parasitic insect.
Nature,Lond., 214, 700.

Rotheram, S. (1973a). The surface of the egg of a parasitic insect.
I. The surface of the egg and first-instar larva of *Nemeritis*.
Proc. Soc. Lond. B, 183, 179-194.

Rotheram, S. (1973b). The surface of the egg of a parasitic insect.
II. The ultrastructure of the particulate coat on the egg of
Nemeritis. *Proc. Soc. of Lond. B*, 183, 195-204.

Rowley, D. and Jenkin, C. R. (1962). Antigenic cross-reaction between host and parasite as a possible cause of pathogenicity.
Nature, Lond., 193, 151-154.

Salt, G. (1956). Experimental studies in insect parasitism. IX.
The reactions of a stick insect to an alien parasite. *Proc.
Roy. Soc. Lond.*, B, 146, 93-108.

Salt, G. (1963a). The defence reaction of insects to metazoan
parasites. *Parasitology*, 53, 527-642.

Salt, G. (1963b). Experimental studies in insect parasitism. XII.
The reactions of six exopterygote insects to an alien
parasite. *J. Insect Physiol.*, 9, 647-669.

Salt, G. (1965). Experimental studies in insect parasitism. XIII.
The haemocytic reaction of a caterpillar to eggs of its
habitual parasite. *Proc. Roy. Soc. Lond. B*, 162, 303-318.

Salt, G. (1966). Experimental studies in insect parasitism. XVI.
The haemocytic reaction of a caterpillar to larvae of its
habitual parasite. *Proc. Roy. Soc. Lond. B*, 165, 155-178.

Salt, G. (1968). The resistance of insect parasitoids to the
defence reactions of their hosts. *Biol. Rev.*, 43, 200-232.

Salt, G. (1970). "The cellular defence reactions of insects."
Cambridge University Press, London.

Salt, G. (1971). Teratocytes as a means of resistance to cellular
defence reactions. *Nature, Lond.*, 232, 639.

Salt, G. (1973). Experimental studies in insect parasitism. XVI.
The mechanism of the resistance of *Nemeritis* to defence
reactions. *Proc. Roy. Soc. Lond.*, 183, 337-350.

Stephens, J. M. (1959). Immune responses of some insects to some
bacterial antigens. *Canad. J. Microbiol.*, 5, 203-228.

Stephens, J. M. (1962). Bactericidal activity of the blood of
actively immunized wax-moth larvae. *Canad. J. Microbiol.*,
8, 491-499.

Stephens, J. M. (1963). Effects of active immunization on total
hemocyte counts of larvae of *Galleria mellonella* (L.).
J. Insect Pathol., 5, 152-156.

Stoltz, D. B., S. B. Vinson, E. A. Machinnon. (1975). Baculovirus-
like particles in the reproductive tracts of female parasitoid
wasps. *Canadian J. Microbiology* (In press).

Streams, F. A. and Greenberg. L. (1969). Inhibition of the
    defense reaction of *Drosophita melanogaster* parasitized
    simultaneously by the wasps *Pseudeucoila bochei* and
    Pseudeucoila mellipes. *J. Invertebr. Path.*, 13, 371–377.
Taylor, R. L. (1969). A suggested role for the polyphenol-
    phenoloxidase system in invertebrate immunity. *J. Invertebr.
    Pathol.*, 14, 427–428.
Vinson, S. B. (1969). Generay morphology of the digestive and
    internal reproductive systems of adult *Cardiochelis nigriceps*
    (Hymenoptera: Branconidae). *Ann. Eetomol. Soc. Am.*, 62,
    1414–1419.
Vinson, S. B. (1970). Development and possible function of
    teratocytes in the host-parasite association. *J. Invertebr.
    Pathol.*, 16, 93–101.
Vinson, S. B. (1971). Defense reaction and hemocytic changes in
    Heliothis virescens in response to its habitual parasitoid
    Cardiochiles nigriceps. *J. Invertebr. Pathol.*, 18, 94–100.
Vinson, S. B. (1972a). Factors involved in successful attack
    on *Heliothis virescens* by the parasitoid *Cardiochiles nigriceps*.
    J. Invertebr. Pathol., 20, 118–123.
Vinson, S. B. (1972b). Effect of the parasitoid, *Campoletis
    sonorensis* on the growth of its host, *Heliothis virescens*.
    J. Insect Physiol., 18, 1501–1516.
Vinson, S. B. (1974a). The role of the foreign surface and female
    parasitoid reactions on the immune response of an insect.
    Parasitology, 68, 27–33.
Vinson, S. B. (1975b). Biochemical coevolution between parasitoids
    and their hosts. *In* "Evolutionary strategies of parasitic
    insects and mites" (Price, P., ed.). pp. 14–48. Plenum Press.
    New York and London.
Vinson, S. B. and D. J. Barras. (1970). Effects of the parasitoid,
    Cardiochiles nigriceps, on the growth, development and tissues
    of *Heliothis virescens*. *J. Insect Physiol.*, 16, 1329–1338.
Vinson, S. B. and W. J. Lewis. (1973). Teratocytes: Growth and
    numbers in the hemocole of *Heliothis virescens* attacked by
    Microplitis croceipes (Cresson). *J. Invertebr. Pathol.*,
    22, 351–355.
Vinson, S. B. and Scott, J. R. (1974). Parasitoid egg shell
    changes in a suitable and unsuitable host. *J. Ultrastruct.
    Res.*, 47, 1–15.
Vinson, S. B. and J. R. Scott. (1975). Particles containing DNA
    associated with the oocyte of an insect parasitoid. *J. Invertebr.
    Pathol.*, 25, 375–378.
Walker, I. (1959). Die Abwehrreaktion des Wirtes *Drosophila
    melanogaster* gegen die Zoophage *Cynipidae pseudeucoila bochie*
    Weld. *Rev. Suisse Zool.*, 66, 569–632.
Walters, D. R. and C. M. Williams. (1966). Reaggregation of insect
    cells as studied by a new method of tissue and organ culture.
    Science, 154, 516–517.

Whitcomb, R. F., Shapiro, M. and Granados, R. R. (1974). Insect
    defense mechanisms against microorganisms and parasitoids.
    In "The Physiology of Insecta." (Rochstein, M., ed.) Vol 5.
    pp. 447-536. Academic Press, New York.

Hemocytes and Phagocytosis in the American Lobster, Homarus americanus

Harriette C. Schapiro

James F. Steenbergen

AND

Zoe A. Fitzgerald

DEPARTMENTS OF BIOLOGY AND MICROBIOLOGY
SAN DIEGO STATE UNIVERSITY
SAN DIEGO, CALIFORNIA

I. INTRODUCTION

Phagocytosis in vertebrates is the initial step in the development of specific immunity against foreign materials. The initial
adherence of a foreign particle to the surface of phagocytes is
usually mediated by opsonins which, in the vertebrates, are
immunoglobulins. The particle is then taken into the phagocyte
where it is usually digested following fusion of the phagosome and
lysosomes. In the invertebrates, a similar series of events takes
place. Since invertebrates lack the specific immunoglobulins
which characterize vertebrate immunity, phagocytosis is much more
significant to their total immunological defense.

Metchnikoff (1893) laid the ground work for our under tanding
of the importance of phagocytosis in defense mechanisms of invertebrates. In *Daphnia*, Metchnikoff described the crucial role of
phagocytosis in defense against *Monospora bicuspidata*. From these
studies, he concluded:

> "The phagocytic action of the leucocytes, so evident
> and easily studied in the transparent Daphniae, destroy
> the spores of pathogenic microbes and prevents their de-
> velopment, thus protecting the invaded organism...If on
> the other hand, the phagocytic action is inadequate, owing
> to the continued increase in number of spores swallowed
> or for any other reason, the latter begin to germinate,
> and give rise to budding conidia...the *Daphnia* succumbs
> in a short time to its attack."

The remainder of the early work on phagocytosis and hemocytes of
invertebrates has been reviewed previously (Huff, 1940; Baer,
1944; Feng, 1967; Bang, 1967; Rabin, 1970; Sindermann, 1971;
Schapiro, 1975).

II. IN VITRO PHAGOCYTOSIS

Cellular defense mechanisms of insects have been reviewed by
Salt (1970). At the time of his review, little work *in vitro*
had been attempted. Since that time, Rabinovitch and De Stefano
(1970) and Scott (1971) have shown *in vitro* phagocytosis of red
cells in the absence of opsonins.

Feng (1967) has reviewed the earlier studies of cellular
responses of molluscs. Recent work with molluscs has emphasized
in vitro phagocytosis. Prowse and Tait (1969) demonstrated that
opsonins are necessary for *in vitro* phagocytosis of foreign
particles by amoebocytes of the snail *Helix aspersa*. Their results indicated a degree of specificity of opsonins for different
classes of foreign materials. Using suspensions of hemocytes in
hemolymph from the oyster, *Crassostrea virginica*, and the clam,
Mercenaria mercenaria, Foley and Cheng (1975) found that all
hemocyte types interact with bacteria. They concluded that one
cell type, the granulocyte, was of greater importance in phago-

cytosis. Bacteria were found adhering to hemocyte surfaces and
in phagosomes. *In vitro* phagocytosis of bacteria by hemocytes of
Mercenaria mercenaria was accompanied by "degranulation" of
hemocytes, with concommitent release of lysozyme (Cheng *et al.*,
1975).

Phagocytosis in decapod crustacea has been studied in detail
by McKay and Jenkin (1970a). They found that *in vitro* phagocytosis
of erythrocytes by hemocytes of the crayfish, *Parachaeraps
bicarinatus*, required specific opsonins. These opsonins appeared
to be hemagglutinins, which enhanced adhesion of erythrocytes to
hemocytes (McKay *et al.*, 1969). They successfully immunized the
crayfish against *Pseudomonas* CP by using an alcohol-killed
vaccine (McKay and Jenkin, 1969). A variety of vaccines from
other gram-negative bacteria or lipopolysaccharide endotoxins
also increased resistance to *Pseudomonas* infection. The immunity
was ascribed to a change in activity of phagocytic cells (McKay
and Jenkin, 1970b, c). It was emphasized that efficient phagocytosis required pretreatment with hemolymph (McKay and Jenkin,
1970c). More recent work has shown that these recognition
factors (opsonins) are present both in the hemolymph and bound to
the plasma membrane of hemocytes (Tyson and Jenkin, 1974).

The American lobster, *Homarus americanus*, represents an ideal
system for the study of phagocytosis. *H. americanus*, like other
crustaceans, can efficiently clear almost all bacteria from their
hemolymph. The notable exception is the disease gaffkemia
(Snieszko and Taylor, 1947), caused by the bacterium *Aerococcus
viridans* (formerly *Gaffkya homari*) (Stewart and Rabin, 1970).
While other bacteria are phagocytized and digested, virulent
A. viridans are phagocytized but apparently not destroyed, and
multiply and eventually kill the host by septicemia (Cornick
and Stewart, 1968).

Paterson and Stewart (1974) utilized a lobster hemocyte
suspension to study *in vitro* phagocytosis of sheep red blood cells.
Very low numbers of hemocytes were phagocytic in this system.
Opsonization of red blood cells increased the yield of phagocytic
hemocytes from 1% to 2% of the total hemocytes in suspension.

III. CURRENT WORK

We have used an *in vitro* system of *H. americanus* hemocytes to
assess relative efficiency of phagocytosis of virulent and avirulent
strains of *A. viridans* (Vandewalker, 1974). In this system, neither
strain of bacteria were successfully phagocytized (2-4% of the
hemocytes contain bacteria) without opsonization. The avirulent
strain, when opsonized, was phagocytized by greater than 90% of
the hemocytes. An unrelated and nonpathogenic bacterium,
Micrococcus luteus, gave similar results. On the other hand, the
virulent strain was phagocytized by approximately 40% of the
hemocytes under the same conditions. Not only were virulent

bacteria taken up by fewer hemocytes, but fewer virulent bacteria were found per phagocytic cell (Table 1).

These results are consistent with observations of Metchnikov (1893) on the role of phagocytosis in disease. He stated that the speed and vigor of phagocytosis early in infection has a marked effect on the outcome of the disease. We have observed that virulent *A. viridans* are phagocytized with relatively poor efficiency. If the bacteria are able to multiply intracellularly, as suggested by observations of Cornick and Stewart (1968), then development of a fatal systemic infection is not surprising.

Paterson and Stewart (1974) report very low levels of phagocytosis of erythrocytes in lobsters compared to other invertebrate systems. We find levels of phagocytosis of bacteria in lobsters comparable to those found by Foley and Cheng (1975) in molluscs. The difference in number of phagocytic hemocytes in our system, and in that of Paterson and Stewart's may be ascribed to the method of preparing hemolymph for opsonization. Paterson and Stewart prepared "serum" for opsonization by withdrawing hemolymph, allowing it to clot, breaking up the clot, and storing it for 24 hr at 4°C. The "serum" was then separated from the clot by centrifugation. Our method involves bleeding a lobster using a cold syringe, transferring the hemolymph to an iced centrifuge tube, and gently removing cells by centrifugation at 4°C. The "plasma" is then held at 4°C and used for opsonization within 3 hr. Paterson and Stewart's prolonged incubation of serum and cells may have caused destruction or inactivation of opsonins. Clotting cells of Crustacea are very fragile (Hearing and Vernick, 1967; Needham, 1970; Stang-Voss, 1971). Clotting of the hemolymph, breaking of the clot, and prolonged incubation undoubtedly leads to release of enzymes which could inactivate opsonins. We have tested this hypothesis by comparing phatocytosis of bacterial cells opsonized with "serum", prepared as described by Paterson and Stewart (1974), and opsonized with "plasma" as prepared by our technique. Opsonization of virulent and avirulent bacteria with "serum" results in approximately half the number of phagocytic cells as found when bacteria are opsonized with "plasma" (Table 2).

Hearing and Vernick (1967) classify lobster hemocytes as eosinophils, ovoid basophils, or spindular basophils. Their identifications were based on electron micrographs and light microscope sections. We have attempted to develop a method for staining and identifying smears of hemocytes. To date our results are not reproducible. As a result, we are not able to differentiate the cell types involved in phagocytosis.

We have been able to immunize the American lobster to gaffkemia using a virulent strain of *A. viridans* (unpubl.). The changes which accompany immunization are unknown. In the crayfish, McKay and Jenkin (1970b, c) have shown that immunization is accompanied by increased phagocytic activity. We are currently investigating phagocytosis by hemocytes isolated from immunized and non-immunized lobsters.

Table 1. Effect of Opsonization on Phagocytosis of Virulent and Avirulent *Aerococcus viridans* by *Homarus americanus* Hemocytes *in vitro*.

Bacteria	Unopsonized		Opsonized	
	% hemocytes showing phagocytosis *	Bacteria per phagocyte +	% hemocytes showing phagocytosis *	Bacteria per phagocyte +
Virulent A. viridans	1.25	2.27 ± 0.64	37.6	4.21 ± 1.55
Avirulent A. viridans	2.5	6.0 ± 1.07	92.0	10.57 ± 2.30
Control Micrococcus luteus	not done	not done	83.0	13.23 ± 2.13

* $\pm$ 3% for average of 3 experiments
+ 95% confidence limits

Table 2. Effects of Opsonizing with "Plasma" or "Serum" (See test).

Bacteria	Opsonized with "Plasma"		Opsonized with "Serum"	
	% hemocytes showing phagocytosis *	Bacteria per phagocyte +	% hemocytes showing phagocytosis ‡	Bacteria per phagocyte +
Virulent Aerococcus viridans	37.6	4.21 ± 1.53	12.0	3.4 ± 1.70
Avirulent Aerococcus viridans	92.0	10.57 ± 2.30	46.0	8.41 ± 1.74

* $\pm$ 3% for average of three experiments
+ 95% confidence limits
‡ Results of one experiment

REFERENCES

Baer, J. G. (1944). Immunité et reactions immunitaires chez les invertebres. Schweiz. Z. *Allg. Pathol. Bact.*, 7, 442–462.

Bang, F. B. (1967). Serological responses among invertebrates other than insects. *Fed. Proc.*, 26, 1680–1684.

Cheng, T. C., G. E. Rodrick, D. A. Foley and S. A. Koehler. (1975). Release of lysozyme from hemolymph cells of *Mercenaria mercenaria* during phagocytosis. *J. Invertebr. Pathol.*, 25, 261–265.

Cornick, J. W., and J. E. Stewart. (1968). Interaction of the pathogen *Gaffkya homari* with natural defense mechanisms of *Homarus americanus*. *J. Fish. Res. Board Can.*, 25, 695–709.

Feng, S. Y. (1967). Responses of molluscs to foreign bodies, with special reference to the oyster. *Fed. Proc.*, 26, 1685–1692.

Foley, D. A., and T. C. Cheng. (1975). A quantitative study of phagocytosis by hemolymph cells of the pelecypods *Crassostrea virginica* and *Mercenaria mercenaria*. *J. Invertebr. Pathol.*, 25, 189–197.

Hearing, V., and S. H. Vernick. (1967). Fine structure of the blood cells of the Lobster, *Homarus americanus*. *Chesapeake Sci.*, 8, 170–186.

Huff, C. G. (1940). Immunity in invertebrates. *Physiol. Rev.*, 20, 68–88.

McKay, D., and C. R. Jenkin (1969). Immunity in the invertebrates. II. Adaptive immunity in the crayfish *(Parachaeraps bicarinatus)*. *Immunology*, 17, 127–137.

McKay, D., C. R. Jenkin and D. Rowley. (1969). Immunity in the invertebrates. I. Studies on the naturally occurring haemagglutinins in the fluid from invertebrates. *Aust. J. Exp. Biol. Med. Sci.*, 47, 125–134.

McKay, D., and C. R. Jenkin. (1970a). Immunity in the invertebrates. The role of serum factors in phagocytosis of erythrocytes by haemocytes of the freshwater crayfish *(Parachaeraps bicarinatus)*. *Aust. J. Exp. Biol. Med. Sci.*, 43, 139–150.

McKay, D., and C. R. Jenkin. (1970b). Immunity in the invertebrates. The fate and distribution of bacteria in normal and immunized crayfish *(Parachaeraps bicarinatus)*. *Aust. J. Exp. Biol., Med. Sci.*, 48, 599–607.

McKay, D., and C. R. Jenkin. (1970c). Immunity in the invertebrates. Correlation of the phagocytic activity of haemocytes with resistance to infection in the crayfish *(Parachaeraps bicarinatus)*. *Aust. J. Exp. Biol. Med. Sci.*, 48, 609–617.

Metchnikoff, E. (1893). Lectures on the comparative pathology of inflammation. Delivered at the Pasteur Institute in 1891. Kegan, Paul, Trench, Trübner and Co. Ltd., London (Republished 1968 by Dover Publications, Inc., New York).

Needham, A. E. (1970). Haemostatic mechanisms in the invertebrata. *Symp. Zool. Soc. London*, 27, 19-44.

Paterson, W. D., and J. E. Stewart. (1974). *In vitro* phagocytosis by hemocytes of the American lobster (*Homarus americanus*). *J. Fish. Res. Board. Can.*, 31, 1051-1056.

Prowse, R. H., and N. N. Tait. (1969). *In vitro* phagocytosis by amoebocytes from the Haemolymph of *Helix aspersa* (Müller) I. Evidence for opsonic factor(s) in serum. *Immunology*. 17, 437-443.

Rabin, H. (1970). Hemocytes, hemolymph, and defense reactions in crustaceans. *Res. J. Reticuloendothel. Soc.*, 7, 195-207.

Rabinovitch, M., and M. DeStefano (1970). Interactions of red cells with phagocytes of the wat moth (*Galleria mellonella* L.) and mouse. *Exptl. Cell Res.*, 59, 272-282.

Salt, G. (1970). "The Cellular Defence Reactions of Insects." Cambridge Monographs in Experimental Biology, No. 16, Cambridge University Press, Cambridge.

Schapiro, H. C. (1975). Immunity in decapod crustaceans. *Amer. Zool.*, 15, 13-19.

Scott, M. T. (1971). Recognition of foreignness in invertebrates. II. *In vitro* studies of cockroach phagocytic haemocytes. *Immunology*, 21, 817-828.

Sindermann, C. J. (1971). Internal defences of crustacea: A review. *Fish. Bull.*, 69, 455-489.

Snieszko, S. F. and Taylor, C. C. (1947). A bacterial disease of the lobster (*Homarus americanus*). *Science*, 105, 500-501.

Stang-Voss, C. (1971). Zur Ultrastruktur der blutzellen wirbelloser Tiere V. Über die hämocyten von *Astacus astacus*. *Z. Zellforsch.*, 122, 68-75.

Stewart, J. E. and Rabin, H. (1970). Gaffkemia, a bacterial disease of lobsters (genus *Homarus*). *In:* "A Symposium on Diseases of Fishes and Shellfishes." (S. F. Snieszko, ed). pp. 431-439. *Am. Fish. Soc. Spec. Publ.* 5, Washington, D. C.

Tyson, C. J., and C. R. Jenkin. (1974). Phagocytosis of bacteria *in vitro* by haemocytes from the crayfish (*Parachaeraps bicarinatus*). *Aust. J. Exp. Biol. Med. Sci.*, 52, 341-348.

Vandewalker, C. J. (1974). A cell culture system for *Homarus americanus* hemocytes. Master thesis, San Diego State University.

Cellular Responses in Decapod Crustaceans to Ascarophis spp. (Spirurida: Nematoda)

GEORGE O. POINAR, JR.

AND

ROBERTA T. HESS

DIVISION OF ENTOMOLOGY AND PARASITOLOGY
UNIVERSITY OF CALIFORNIA
BERKELEY, CALIFORNIA

I. INTRODUCTION

Of the spirurid nematodes known to utilize decapods as
intermediate hosts, only members of the genus *Ascarophis* have
been reported to elicit cellular host responses (Poinar and
Kuris, 1975; Uzmann, 1967). As in spirurid parasites of insects,
this response is an attempt by the crustacean to wall off the
parasite. Although the nematode may be surrounded by host
tissue, the response generally has little overall effect since
spirurids can reach the infective stage and survive quite well
within host capsules.

While studying nematodes associated with marine decapods, two
major types of cellular host responses were encountered. These
responses were studied in detail and compared with those
reported in insects. The results illustrate that some marine
decapods are routinely capable of encapsulating and perhaps
destroying spirurid namatodes.

II. MATERIALS AND METHODS

Specimens of the ghost shrimp, *Callianassa californiensis;*
the shore crab, *Hemigrapsus oregonensis;* and the hermit crab,
Pagurus samuelis, were collected from Bodega Bay and Shell Beach,
California, from 1972-1975. The crustaceans were dissected in
sea water, and host capsules containing nematodes were removed
and examined under the light microscope. For electron microscopy, capsules containing nematodes were fixed in 4% glutaraldehyde in Millonig's phosphate buffer with 0.25 M sucrose for 1 hr,
then transferred to a 1% buffered solution of osmium tetroxide
for 1 hr at 4°C. Following fixation, the capsules were rinsed in
buffer, dehydrated in an alcohol series, and embedded in Araldite
6005. Sections made with glass knives mounted in a Porter-Blum
MT-2 microtome were stained with saturated aqueous uranyl acetate
followed by lead citrate and examined with an RCA-3F and a
Philips EM-300 electron microscope.

III. RESULTS

Host capsules containing *Ascarophis* spp. were removed from the
hemocoel of *C. californiensis, H. oregonensis,* and *P. samuelis.*
Capsule shape varied from spherical to elliptical, ranging in
size from 0.5 to 2.5 mm. In *C. californiensis,* the thick granular
capsules were always attached to the muscle bundles surrounding
the pyloric stomach. They often protruded into the hemocoel,
but remained attached to the pyloric stomach by a narrow band
of tissue. The nematodes rested in cavities filled with a fluid
arising from the breakdown of the adjacent cells (Fig. 1). In
H. oregonensis, nematode capsules occurred throughout the hemocoel
of the crab and exhibited a range in structure. Those found in
the appendages and thorax were thin and fragile (Fig. 2), whereas

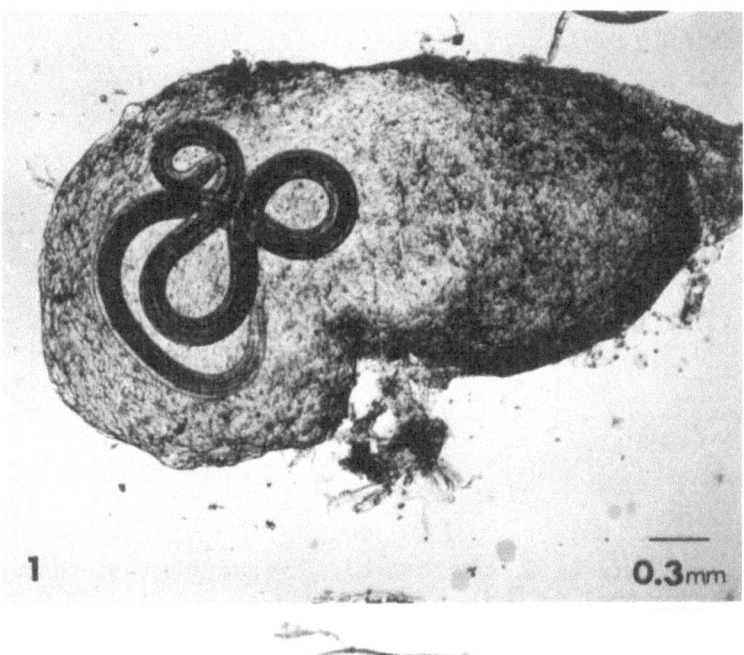

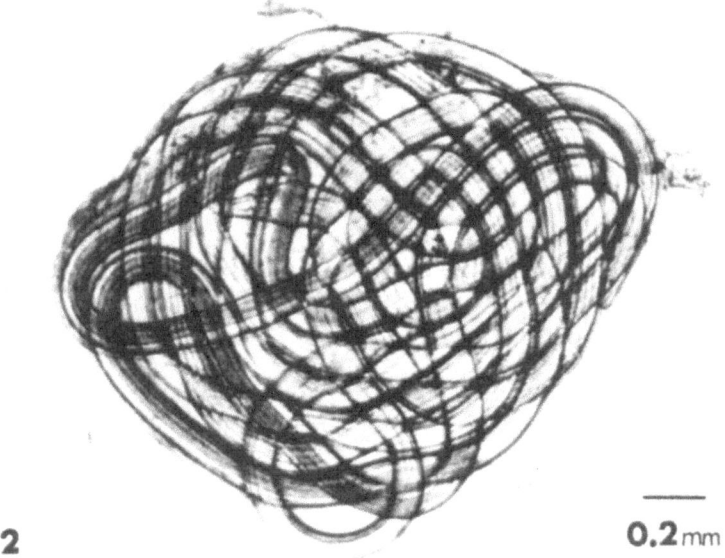

Fig. 1. *Ascarophis* enclosed in a muscular capsule from *Callianassa californiensis*.

Fig. 2. *Ascarophis* enclosed in a membraneous capsule from *Hemigrapsus oregonensis*.

Fig. 3.   Hemocyte of *Callianassa califoriensis* occasionally en-
          countered in the muscular capsules surrounding *Ascarophis*.
          N = nucleus   I = inclusion granules containing microtubules.
          Bar equals 1 μm.

Fig. 4.   Hemocyte of *Hemigrapsus oregonensis* involved in the for-
          mation of capsules surrounding *Ascarophis*.   N = nucleus
          I = inclusion granules.   Bar equals 1 μm.

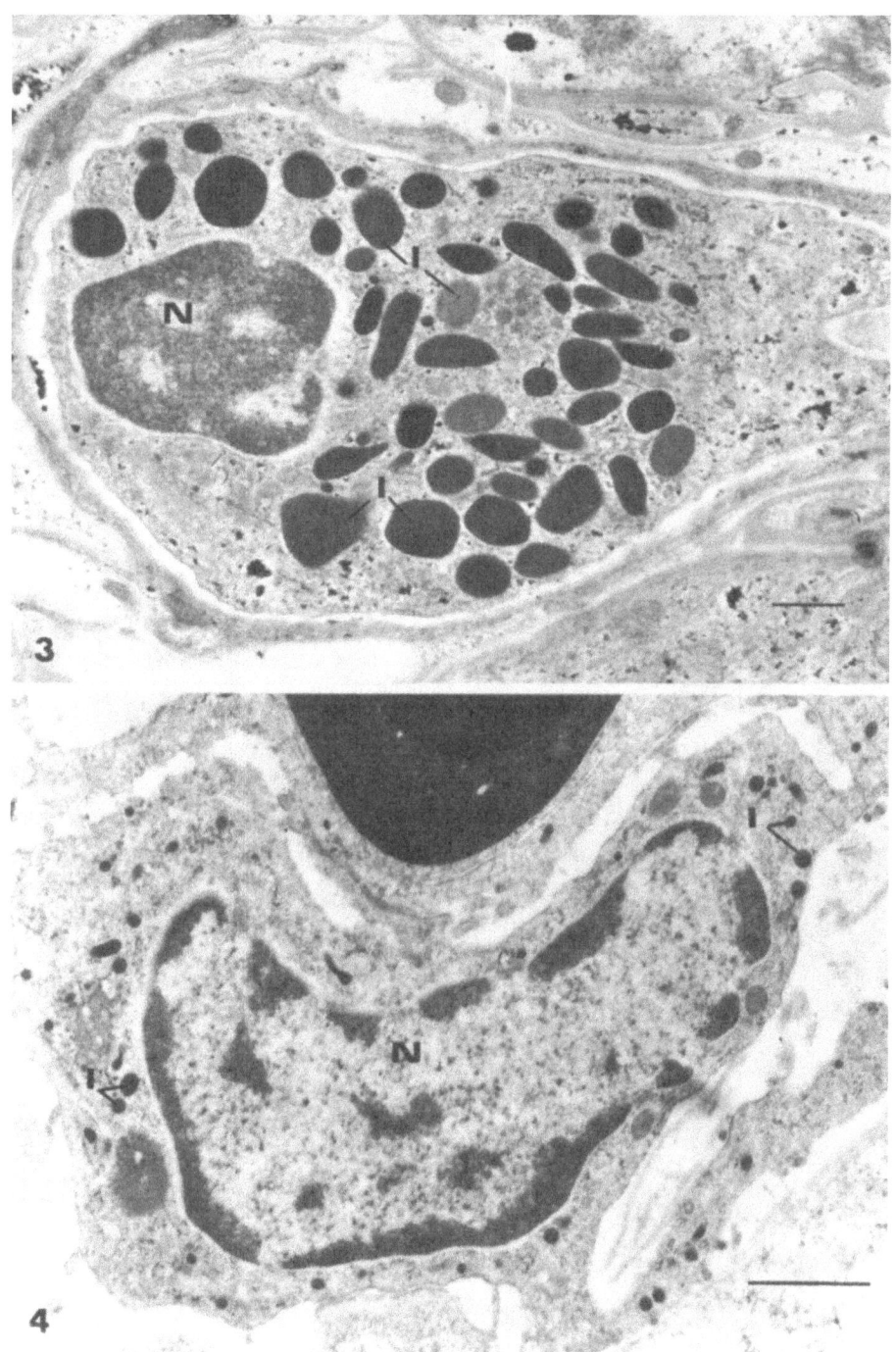

Fig. 5. Section of a muscular capsule surrounding *Ascarophis* in
         Callianassa californiensis. A = nematode C = capsular
         fluid R = residual deposit D = electron-dense region
         M = muscle cells. Bar equals 1 μm.

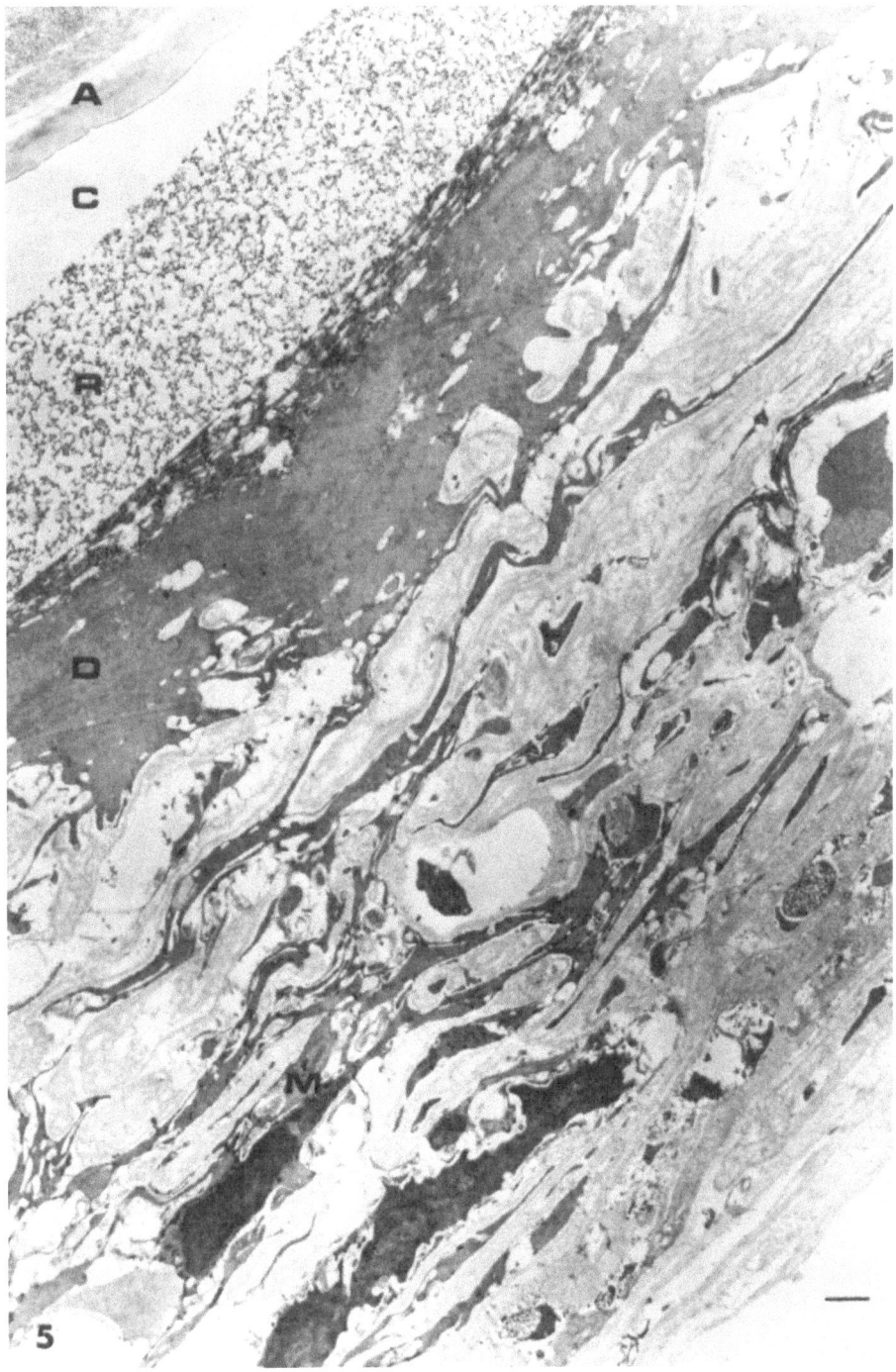

Fig. 6. Section of a young capsule surrounding *Ascarophis* in
         Callianassa californiensis showing region of degenerating
         muscle cells (G) adjacent to the parasite. M = muscle
         cells. Bar equals 1 μm.

Fig. 7. Section of an older capsule surrounding *Ascarophis* in
         Callianassa californiensis showing residual deposit (R)
         of degenerated muscle cells. M = muscle cell. Bar
         equals 1 μm.

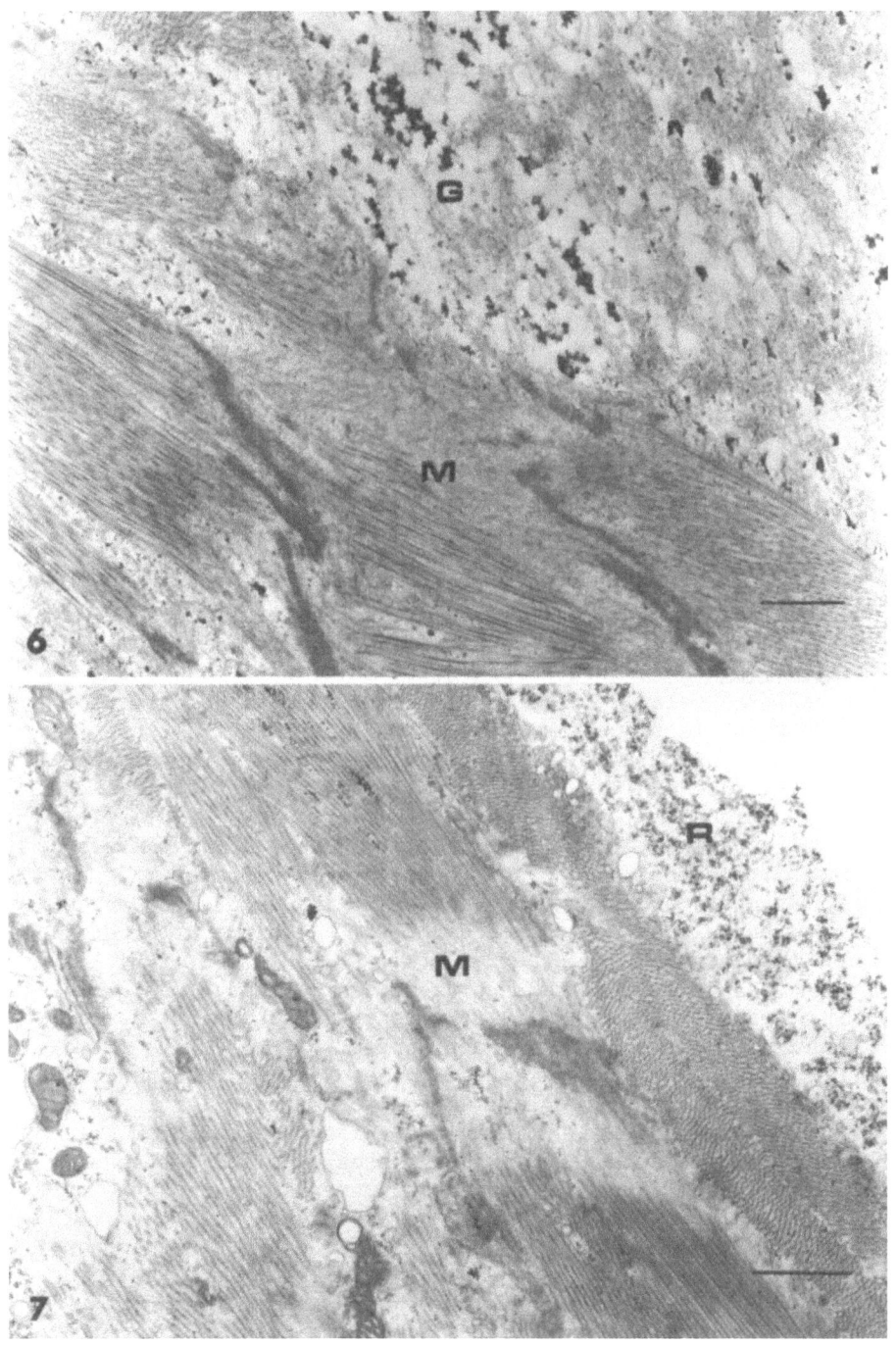

those associated with the hepatopancreas were thick and granular.
Capsules intermediate between these two types were associated
with the pyloric stomach, midgut, hindgut, gonads, and carapace.
In *P. samuelis*, the capsules varied from membraneous to granular
and occurred in the abdomen. Some were attached to the epithelium
lining the abdominal wall, while others were found adjacent to the
huge abdominal muscle bundle.

Descriptions of the nematodes in the above mentioned decapods,
including data on the incidence of infection and seasonal distribution are given elsewhere (Poinar and Kuris, 1975; Poinar and
Thomas, 1976). The definitive hosts, and thus the adult stages
of these nematodes, are unknown.

Capsules removed from the pyloric stomach of *C. californiensis*
were examined with the electron microscope. The capsular walls
were composed of layers of muscle cells surrounding the enclosed
parasites (Fig. 5). The nematodes were lying in fluid-filled
cavities which were lined with a residual deposit of cellular
debris from destroyed host cells.

In younger capsules, an indistinct zone existed between degenerating muscle cells adjacent to the nematode and healthy
muscle cells (Fig. 6). In older capsules all that remained of the
degenerating muscle cells was a residual deposit (Fig. 7).

Outside the residual deposit in mature capsules was an electron-
dense region of necrotic, decomposed muscle cells (Fig. 5). This
was followed by the remainder of the capsular wall comprising
living cells as well as those in various stages of degeneration.
Hemocytes with electron-dense inclusion bodies (Fig. 3) were
present in the hemocoel of *C. callianassa*; however, they were only
occasionally found in the capsular wall and were never considered
essential elements in capsule formation.

On two occasions, juvenile *Ascarophis* surrounded by layers of
Callianassa hemocytes were found in the host's hemocoel (Fig. 8).
These individuals probably broke out of their muscular capsules
and were encapsulated by host blood cells. A thin enucleate
membrane bordered the parasites and their capsules (Fig. 9).
These capsules were pigmented and rigid, and although the enclosed
nematodes were still living, it is doubtful whether they could
have survived these conditions much longer.

Electron microscopical examinations of *Ascarophis* capsules in
P. samuelis revealed a muscular type of formation (Fig. 10). The
capsular wall was composed of a fairly uniform series of muscle
cells containing elongate nuclei and masses of fibers. There
were no large regions of electron-dense material or necrotic cells
in these capsules.

The capsules of *H. oregonensis* were of two types: those composed of muscle tissue and those formed by hemocytes. The former
capsules were very similar to those reported above in
Callianassa and *Pagurus* with layers of muscle cells. Young
capsules contained degenerating muscle cells in the fluid-filled

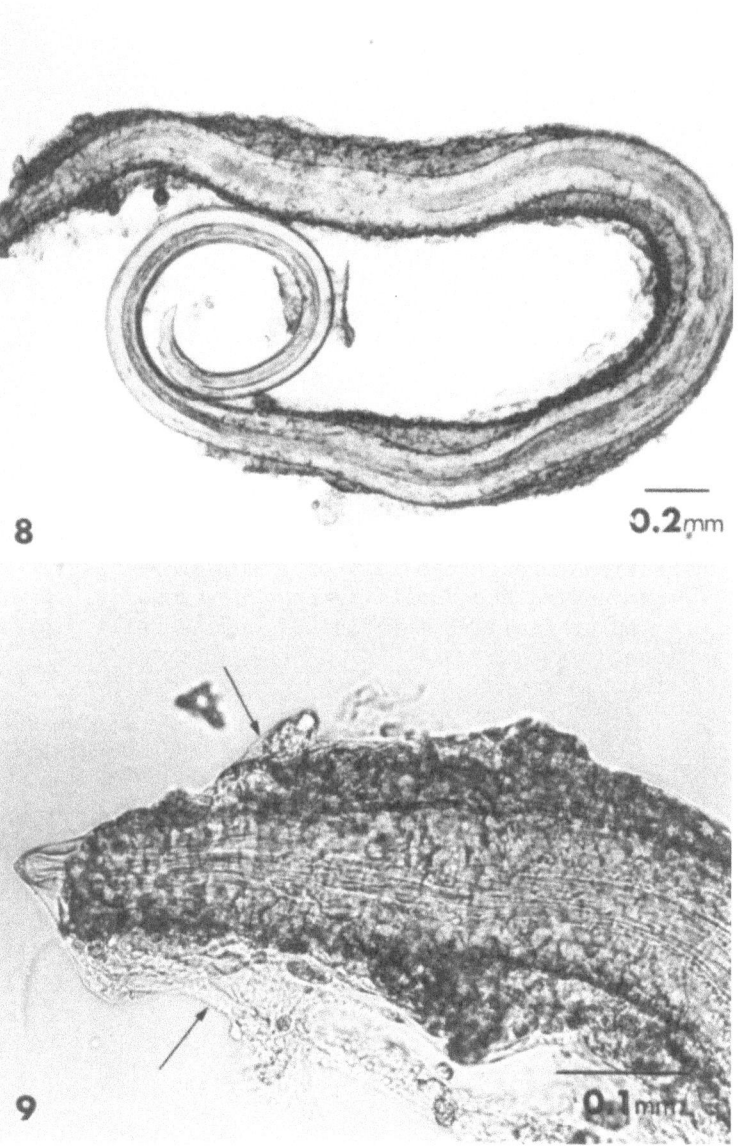

Fig. 8. Hemocytic encapsulation of *Ascarophis* in *Callianassa californiensis*. Note rigid, pigmented capsule.

Fig. 9. Anterior portion of encapsulated *Ascarophis Callianassa californiensis*. Note thin enucleate membrane bordering parasite and capsule (arrows).

Fig. 10. Section through the wall of a muscular capsule
          surrounding *Ascarophis* in *Pagurus samuelis*. C =
          capsular fluid N = nuclei of muscle cells. Bar equals
          1 μm.

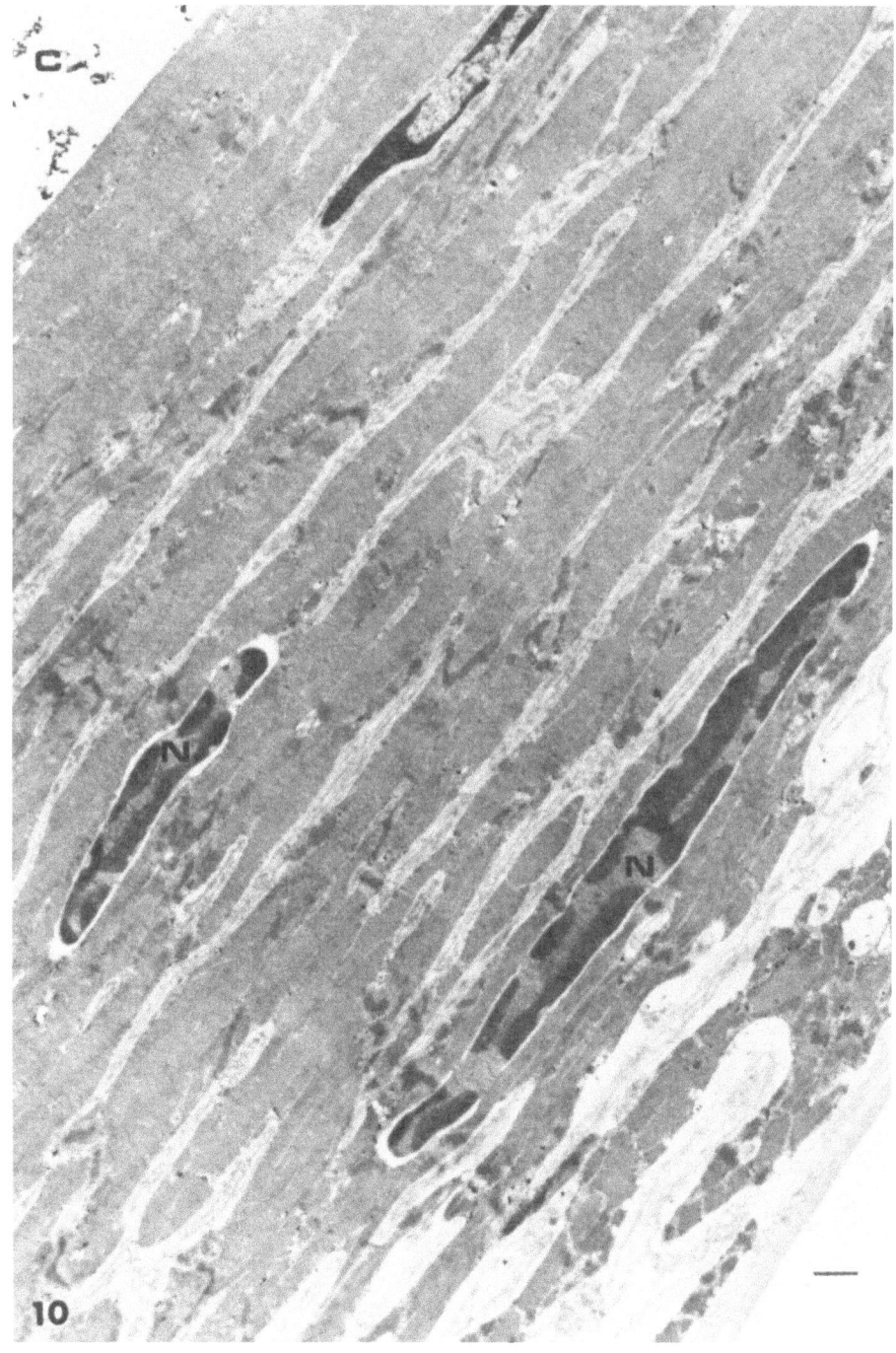

Fig. 11.   Section through a young muscular capsule surrounding
           Ascarophis in *Hemigrapsus oregonensis* showing degener-
           ation of muscle cells in the nematode's cavity (G).
           M = muscle cells. Bar equals 1 μm.

Fig. 12.   Section through an older muscular capsule surrounding
           Ascarophis in *Hemigrapsus oregonensis* showing the residue
           deposit (R) from degenerated cells and zone of healthy
           muscle cells (M). C = nematode's cavity. Bar equals
           1 μm.

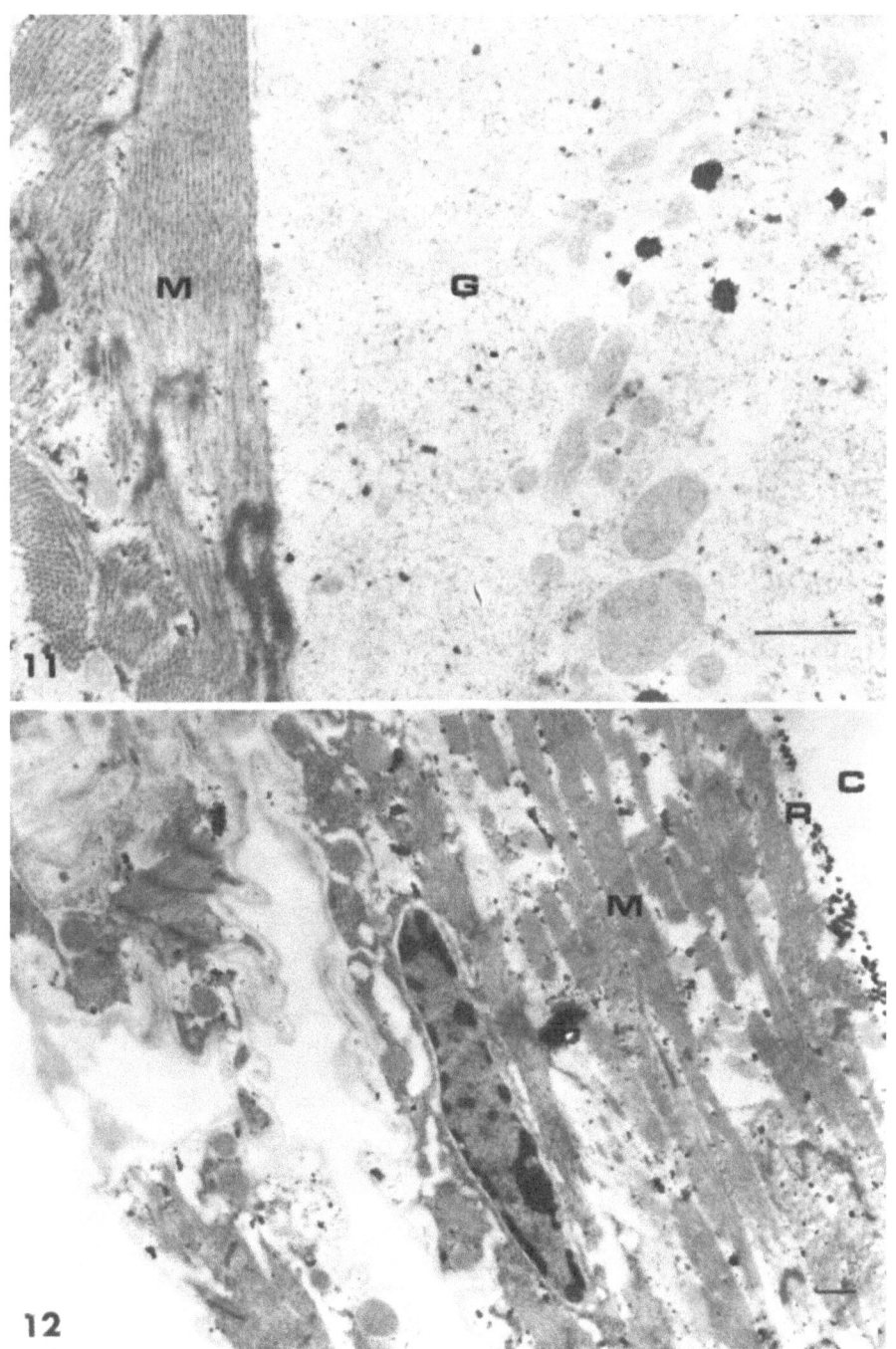

Fig. 13. Section through the wall of a hemocytic capsule
          surrounding *Ascarophis* in *Hemigrapsus oregonensis*.
          Note inner electron-dense layer (arrows) and outer
          region of flattened blood cells. C = nematode's
          cavity. Bar equals 1 μm.

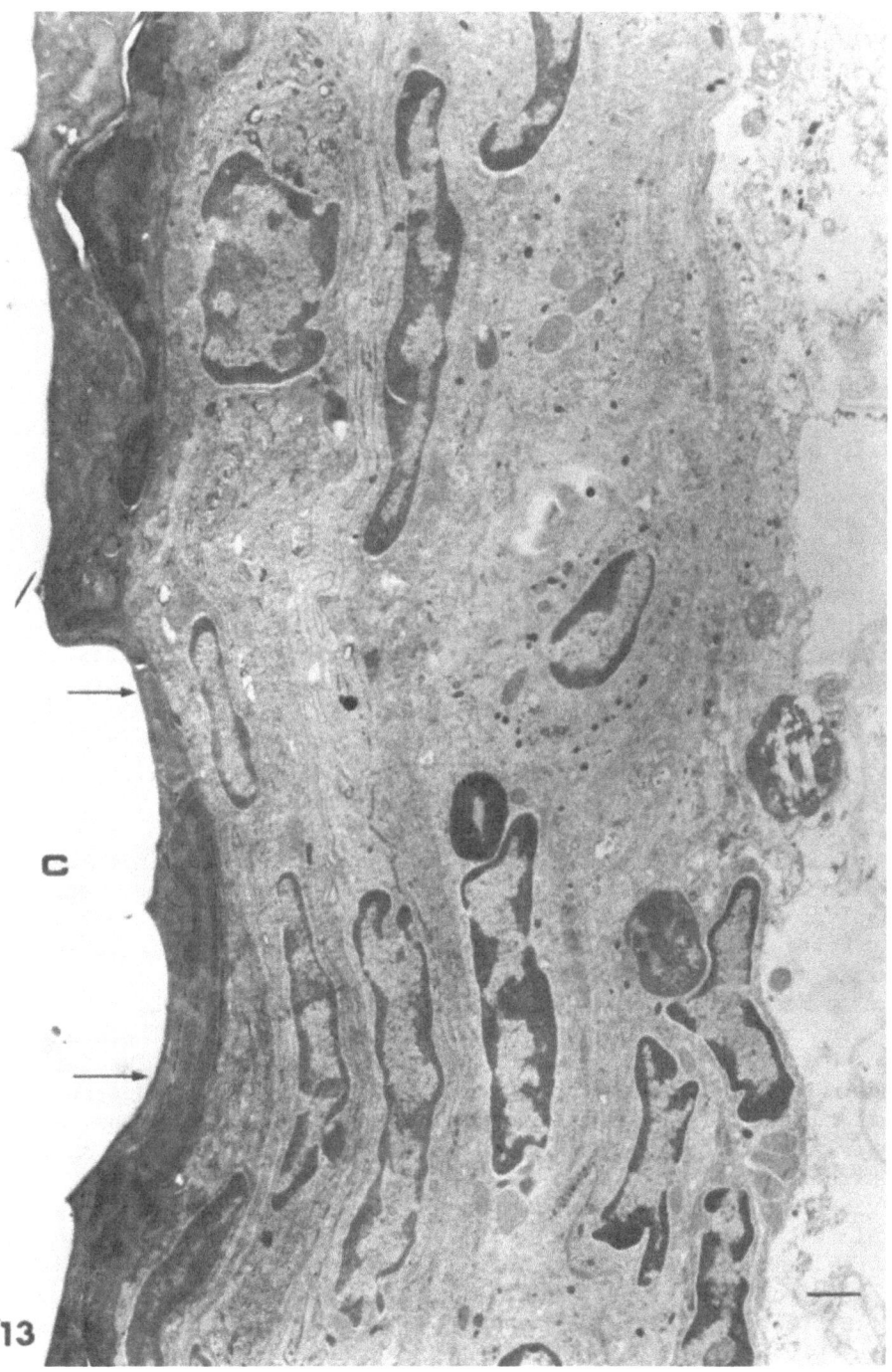

cavities (Fig. 11), while older capsules possessed an outer
residual deposit (Fig. 12).

The walls of the hemocytic capsules were composed of five or
more layers of blood cells (Fig. 13). The blood cells of
Hemigrapsus taking part in this reaction were similar to insect
plasmatocytes (Fig. 4). In *Hemigrapsus*, these hemocytes were
variable in shape, but generally spherical to elliptical. Their
surface was irregular, and their cytoplasm contained mitochondria
and small inclusion granules. The hemocytes forming the capsule
were flattened, and the inner layer had modified into an electron
dense layer. This layer was closely appressed to the nematode, and
its original plastic nature could be observed by retention of the
cuticular pattern of the nematode (Fig. 13). Just outside the
electron-dense region was a layer of darkened necrotic cells,
followed by zone of flattened but normal appearing hemocytes.
Normal appearing spherical blood cells were found on the
periphery of the capsule.

IV. DISCUSSION

Little is known about the defense reactions of decapods to
metazoan parasites. By inserting cellodian tubes in the hemocoel
of *Potamobius leptodactylus*, Danini (1955) showed a hemocytic
encapsulation response in crabs. In studying nematode parasites
of *Pagurus*, Díaz (1969) did not mention the presence of host
capsules and Feigenbaum (1973) definitely stated that
Ascarophis spp., as well as other nematodes, were not encapsulated in the hemocoel of *Penaeus* shrimp. Only Uzmann (1967)
discovered *Ascarophis* juveniles in lenticular cysts on the rectal
wall of the lobster, *Homarus americanus*.

In the present investigation on capsule formation in three
marine decapods, two major types of responses were noted. One
involved muscle tissue, the other blood cells. The muscle response
was noted in all three hosts and consisted of concentric layers of
muscle cells surrounding the parasite.

In *Callianassa*, the parasite encounters muscle bundles as it
passes toward the hemocoel from the pyloric stomach. There are
two possible explanations for the resulting capsules. The nematode may actually seek out muscle tissue. This would suggest
that the muscle cells were passive elements in the reaction, and
their response would be similar to that of simple injury. Another
possibility is that the muscle cells actively respond to the
parasite by possibly shifting their position somewhat, which would
explain the presence of elliptical capsules that protrude into the
host's hemocoel, and electron-dense deposits at the periphery of
the nematode's cavity. This matter can only be resolved when
capsule formation is investigated. Nevertheless, the parasite,
along with a surrounding "envelope" of muscle tissue, is forced
out of the muscle bundle and remains attached by only a small

piece of connecting tissue. This reaction appears similar to that which occurs in the roach, *Blattella*, against the invading spirurid nematode, *Physaloptera turgida*. Alicata (1935) found that the nematodes penetrated into host muscles, and the resulting capsule was "sometimes pushed out into the body cavity until its attachment to the muscle was merely by a thin strand."

Results from the present investigation show that degeneration of the surrounding muscle cells is often accompanied by the formation of an electron-dense layer similar in many respects to melanin. Chemical tests to determine the presence of melanin in these capsules were not made, however.

The second type of host response, hemocytic encapsulation, occurred in *Callianassa* and *Hemigrapsus*. However, this response in *Callianassa* was distinct from that which occurred in *Hemigrapsus*. In *Callianassa*, the response was highly successful since the parasite was encased in a rigid coat which restricted movement and may have killed the parasite. In *Hemigrapsus*, the nematodes were surviving within the capsules and presumably were infective to susceptible vertebrates.

The reaction in *Hemigrapsus* was very similar to examples of melanotic encapsulation in insects (Poinar *et al.*, 1968). The blood cells involved in both cases were similar in appearance and function. Those involved in *Hemigrapsus* resembled the plasmatocytes of *Diabrotica* and, using the scheme devised by Stang-Voss (1971) for *Astacus astacus*, would fall into the category of phagocytic amoebocytes.

The electron-dense material found on the inside of the hemocytic capsules in *Hemigrapsus* was probably melanin although diagnostic tests were not performed. However, its location, the surrounding layer of necrotic cells, and the flattened hemocytes making up the majority of the capsule was the same basic structure as found in melanotic capsules in *Diabrotica* (Poinar *et al.*, 1968).

In studying the nematode, *Acanthocheilus quadridentatus*, in *Pagurus prideauxi*, Diaz (1969) stated that occasionally nematodes found in this host had become hard, granular, and darkened. He noted that this reaction could be lethal.

It is now clear that some decapod crustaceans have a cellular defense system, similar to insects. Not only can their hemocytes react to metazoan parasites, but their muscle cells also may respond to the presence of parasitic nematodes. Under certain conditions, the hemocytic response may be effective enough to halt further development or even destroy the parasite.

SUMMARY

Cellular responses to *Ascarophis* nematode infections in *Callianassa californiensis*, *Hemigrapsus oregonensis*, and *Pagurus samuelis* were examined with the electron microscope. In all

three hosts, nematodes were found enclosed in capsules formed
by muscle cells. In *C. californiensis* and *H. oregonensis,*
nematodes were also encapsulated by host blood cells. In the
former host, these hemocytic responses may destroy the parasite
and thus prove successful immune reactions. In the latter host,
the parasites were able to develop to the infective stage and
remain viable within the hemocytic capsules.

REFERENCES

Alicata, J. E.(1935). Early developmental stages of nematodes
    occurring in swine. *U.S.D.A. Tech. Bull.,* <u>489</u>, 1-96.
Danini, E. S. (1925). Beiträge zur vergleichenden Histologie des
    Blutes und des Bindegewebes III. Über die entzündliche
    Bindegewebsneubildung beim Flusskrebs (Potamobius leptodactylus)
    Z. Mikr. Anat. Forsch., <u>3</u>, 558-608.
Diaz, J. P. (1969). Contribution a l'étude de quelques nématodes
    de crustacés. Thesis, University of Montpellier, France.
Feigenbaum, D. L. (1973). Parasites of the commercial shrimp,
    Penaeus vannamei Boone and *Penaeus brasiliensis* Latreille.
    M.S. Thesis, University of Miami, Coral Gables, Florida.
Poinar, Jr. G. O. and Kuris, A. M. (1975). Juvenile *Ascarophis*
    (Spirurida: Nematoda) parasitizing intertidal decapod
    crustacea in California with notes on prevalence and effects
    on host growth and survival. *J. Invertebr. Path.,* <u>26</u>, 375-382.
Poinar, Jr., G. O., Leutenegger, R. and Götz, P. (1968). Ultra-
    structure of the formation of a melanotic capsule in
    Diabrotica (Coleoptera) in response to a parasitic nematode
    (Mermithidae). *J. Ultrastr. Res.,* <u>25</u>, 293-306.
Poinar, Jr., G. O., and Thomas, G. M. (1976). Occurrence of
    Ascarophis (Nematoda: Spirurida) in *Callianassa californiensis*
    (Dana) and other decapod crustaceans. *Proc. Helm. Soc. Wash.,*
    <u>43</u>, 28-33.
Stang-Voss, C. (1971). Zur Ultrastruktur der Blutzellen wirbel-
    loser Tiere. V. Über die Hämocyten von *Astacus astacus* (L.)
    (Crustaces). *Z. Zellforsch.,* <u>122</u>, 68-75.
Uzmann, J. R. (1967). Juvenile *Ascarophis* (Nematoda: Spiruroidea)
    in the American lobster, *Homarus americanus. J. Parasitol.,*
    <u>53</u>, 218.

Comparative Ultrastructural Studies of Cellular Immune Reactions and Tumorigenesis in Drosophila

A. J. NAPPI

DEPARTMENT OF BIOLOGICAL SCIENCES
STATE UNIVERSITY OF NEW YORK
OSWEGO, NEW YORK

155

I. INTRODUCTION

Cellular immunity in insects is mediated by blood cells, or hemocytes, which eliminate invading foreign organisms by phagocytosis and encapsulation (Salt, 1963, 1970; Poinar, 1969, 1974; Whitcomb *et al.*, 1974; Nappi, 1974, 1975a). Encapsulation reactions are typically characterized by the aggregation and adhesion of hemocytes forming multilayered capsules which melanize around foreign surfaces too large to be engulfed by a single host cell. Although many significant contributions have been made to the study of insect immunity, important questions concerning the mechanism of hemocyte activation and the specificity of cellular interactions with various infectious and oncogenic agents still remain unanswered. An elementary understanding of how immuno-competent cells differentiate between 'self' and 'not-self' and react to eliminate foreignness necessitates a working knowledge of the origin and diverse functions of specific cell types, and of the mechanisms which regulate their activity throughout development and under various pathological conditions. Unfortunately, little of this information is available for insects, and only for a very few species.

In larvae of *Drosophila* the encapsulation reaction made against certain metazoan parasites (Walker, 1959; Nappi and Streams, 1969; Nappi, 1975b) is believed to be similar to the reaction which forms abnormal cellular growths, termed melanotic tumors, in the body cavity of various mutant strains of the fly (Oftedal, 1953; Castiglioni, 1957; Rizki, 1957, 1960; Perotti and Bairati, 1968; Rizki and Rizki, 1974a,b). In both cases the hemocytic reactions involve a precocious mass differentiation of hemocytes which aggregate and adhere to one another to form pigmented capsules. This report compares the pathological changes occurring during melanotic tumor formation in some mutant strains of *D. melanogaster* with those changes taking place during the encapsulation of the wasp parasite, *Pseudeucoila bochei*. The purpose of the study was to determine whether there were features common to both processes which might provide information about hemocyte specificity and the recognition of foreignness.

II. MELANOTIC TUMOR FORMATION

Melanotic tumors in *Drosophila* have been known for a long time. The development of these genetically controlled masses has been analyzed by numerous workers, and it has been demonstrated that several distinct types exist. The few histological studies that have been done have indicated that melanotic tumors are formed principally by the aggregation of hemocytes among various tissues and organs, and the melanization of the encapsulated masses (see Scharrer and Lochhead, 1950; Harshbarger and Taylor, 1968; Ghelelovitch, 1969). The majority of tumors are benign, and the

cells show little or no cell division. The intra- and intercellular melanization may become so extensive that the nature and
origin of the melanotic masses from the hemocytes is frequently
obscured. The hemocytes chiefly involved in tumorigenesis are
lamellocytes, which are large and flattened cells that differentiate from the predominant blood cell type, the plasmatocyte
(Oftedal, 1953; Rizki, 1957, 1960). The tranformation of
plasmatocytes to lamellocytes is believed to be under hormonal
control, occurring in non-tumorous larvae at about the time of
pupation (Rizki, 1960). The manifestation of the tumor phenotype
is controlled by a complex genotype. Depending on the strain
and rearing conditions, the incidence of tumors varies consider-
ably, as does the size, shape, number, time of appearance, and
location of the masses within the body.

The first investigator to describe the development of a
hereditary tumor, "lethal 7", in *Drosophila* was Stark (1918). The
tumors were believed to develop in certain ganglia, the salivary
glands, and in imaginal discs located near the posterior end of the
larval abdomen. Stark considered the melanotic masses to be
abnormal cellular overgrowths, lethal to the hosts, and "somewhat
resembling the tumors of vertebrates." She found that in the
early stages of tumor development there was first a deposit of
pigment, followed by a rapid proliferation of cells. The cells
near the center of the tumor were large, polyhedral, spheroidal,
or fusiform in shape. As development of the tumor progressed the
older cells, filled with pigment, were pushed toward the periphery
and became flattened and closely crowded together "forming
laminated layers of pigment giving the tumor the appearance of
being encapsulated." However, a study of her published figure of
a tumor arising in the posterior end of the larval body (Fig. 1)
suggests the growth represents host tissue encapsulated by
hemocytes.

It is interesting that extirpation of the "lethal 7" tumors
prolonged the life of the larvae, but the operation did not allow
the insects to reach the pupal stage and they died as usual
(Stark, 1918). Implants of the lethal tumor into normal larvae
caused the death of all hosts before pupation. However, in a
nonlethal mutant of the same strain, Stark (1919) found that tumors
implanted into host larvae could be carried into the adult stage.
Moreover, some larvae from which tumors had been removed were
able to develop to the adult stage and to eventually produce off-
spring with tumors.

Unfortunately, very little information is available concerning
the fine structure of melanotic tumors in *Drosophila*. Bairati
and Perotti (1966) and Perotti and Bairati (1968) studied the
ultrastructure of tumors in two mutant strains, tuB_3 and *Frackled*.
In the tuB_3 strain, the melanotic masses were composed of either
small aggregates of hemocytes or detached fragments of the hemolymph organ (= lymph glands or "blood-forming organs") encapsulated

by large flattened hemocytes (lamellocytes). During the early
stages of tumorigenesis, electron-opaque material, identified as
melanin, was gradually deposited, first in the intercellular
spaces along the cell membranes, and later within the encapsulated cells so that the aggregations became completely melanized.
The hemocytes surrounding the melanized portions of the capsule
were arranged in two or three layers. The cells were greatly
extended at their surfaces and possessed fingerlike projections
interdigitating with those of adjacent cells forming a continuous,
compact sheet. The cytoplasm contained numerous inclusions with
a myelin configuration, poorly developed endoplasmic reticulum,
groups of free ribosomes, and very few mitochondria. There was
no evidence of specialized, pigment containing organelles similar
to vertebrate melanosomes in the cells undergoing melanization.

During the early stages of tumor formation in the mutant
strain *Freckled*, melanization occurred intracellularly, arising
first in the perinuclear region of pupal fat body cells. During
pupal development there was a progressive deposition of fine
particulate material (10-15 Å), and in the adult stage the nuclei
and cytoplasm of the fat body cells were completely masked by
a dense deposit of pigment.

Rizki and Rizki (1974a,b) examined by means of scanning
electron microscopy the topology of the caudal fat body of the
mutant strain tu^W. Portions of this tissue became encapsulated
by host hemocytes shortly before pupation. In forming the capsule
spherical plasmatocytes were precociously transformed into large,
extremely flattened lamellocytes which adhered to the fat body
cells producing compact masses that later melanized. The malanotic
masses were retained within the body of the fly throughout adult
life. The first evidence of pathological change was the disintegration of the basement membrane surrounding the larval fat body
cells. In addition, small globular particles or droplets, which
resembled the cytoplasmic inclusions of fat body cells, were found
between and on the surfaces of the dissociating fat cells (Figs. 2
-5).

In larvae of the tumor strain $vg\ tu^{28}$, the pigmented masses
which formed in various regions of the body cavity were found to be
comprised of encapsulated and malanized fat body cells (Nappi,
1975c). Preliminary studies with the light microscope (unpubl.)
showed the early stages in the formation of tumors were characterized
by the aggregation of large numbers of hemocytes, both lamellocytes
and plasmatocytes, around certain fat body cells, some of which
had dissociated. The plasma membrane and peripheral regions of the
fat cells were the first to melanize (Figs. 6,7). Later, entire
intact fat body cells and the components of ruptured cells became
heavily pigmented.

At the ultrastructure level early evidence of pathological
changes within affected fat cells consisted of the formation of
concentrically lamellated cytoplasmic inclusions, i.e., lamellar

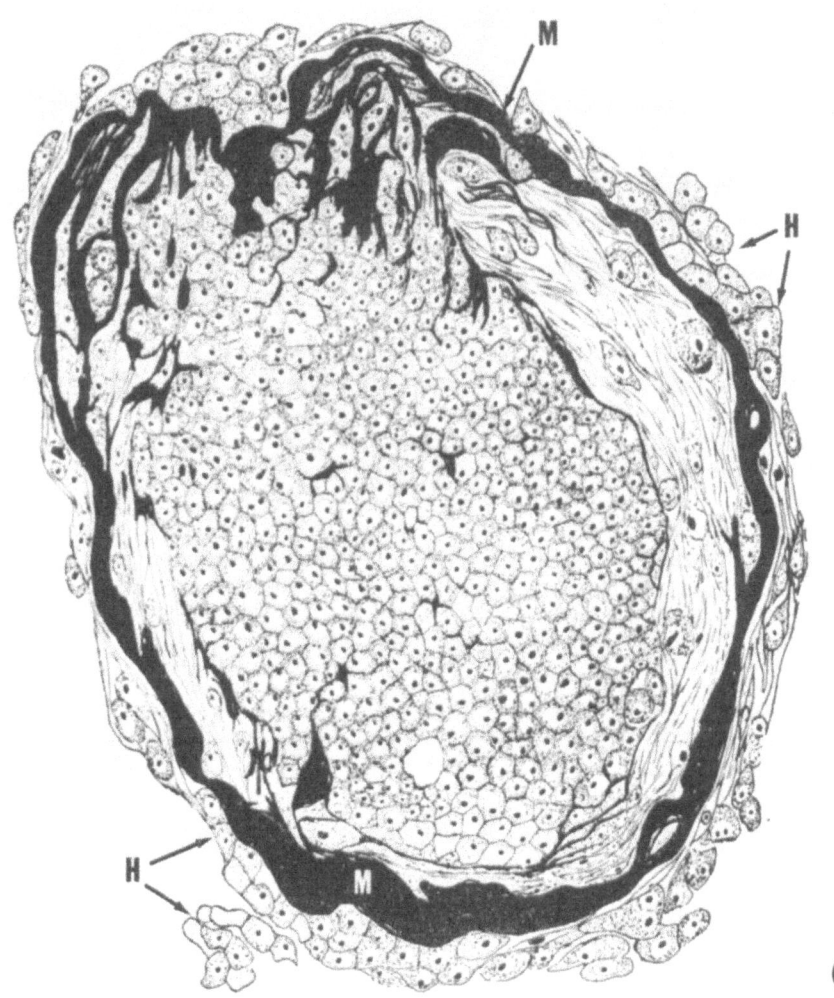

Fig. 1.   Melanotic tumor developing in embryonic cells near the
          posterior end of "lethal 7" mutant larva of
          Drosophila melanogaster.   M, melanin; H, cells believed
          to be hemocytes. (From Stark, 1918).

Fig. 2.   Scanning electron micrograph of noraal surface features
          of caudal fat body of non-tumorous larva 72 hours of age.
          Note continuity of basement membrane over the surface
          and distribution of crater-like depressions. X 2700.
          Scale, 10μm. (From Rizki and Rizki, 1974a. Courtesy of
          authors).

Fig. 3.   Fat body of 72 hour old tu^w larva. Note the disinte-
          gration of the basement membrane covering the fat body
          cells and the dissociation of the individual cells. X
          2700. Scale, 10 μm. (From Rizki and Rizki, 1974a.
          Courtesy of authors).

Fig. 4.   Fat body of 86 hour old tu^w larva. Surface of cell in
          lower left of frame has been modified by hemocyte (arrow)
          activity. Note normal surface features of the two fat
          cells above. X 3000. Scale, 10 μm. (From Rizki and
          Rizki, 1974b. Courtesy of authors).

Fig. 5.   Enlargement of hemocyte shown in Figure 4. The cell is
          at a stage in the transformation to a lamellocyte. X
          10000. Scale, 2 μm. (From Rizki and Rizki, 1974b.
          Courtesy of authors).

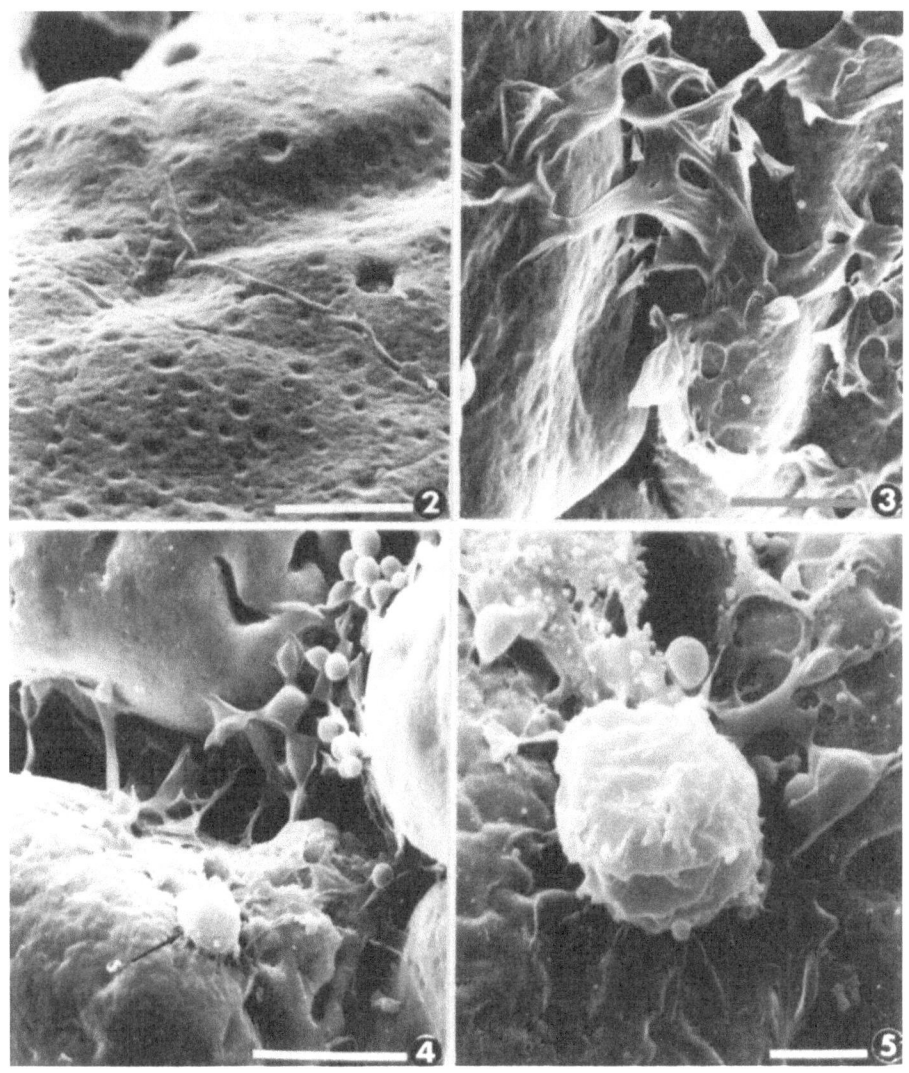

Fig. 6.   Light micrograph showing accumulation of a large mass of
          hemocytes (H) among abnormal fat body cells (F).
          Scale, 30 μm.

Fig. 7.   A different serial section of the same region as that
          shown in Figure 6.   Note envelopment by hemocytes (H),
          and melanization (M) of peripheral areas of fat body.
          Scale, 30 μm.

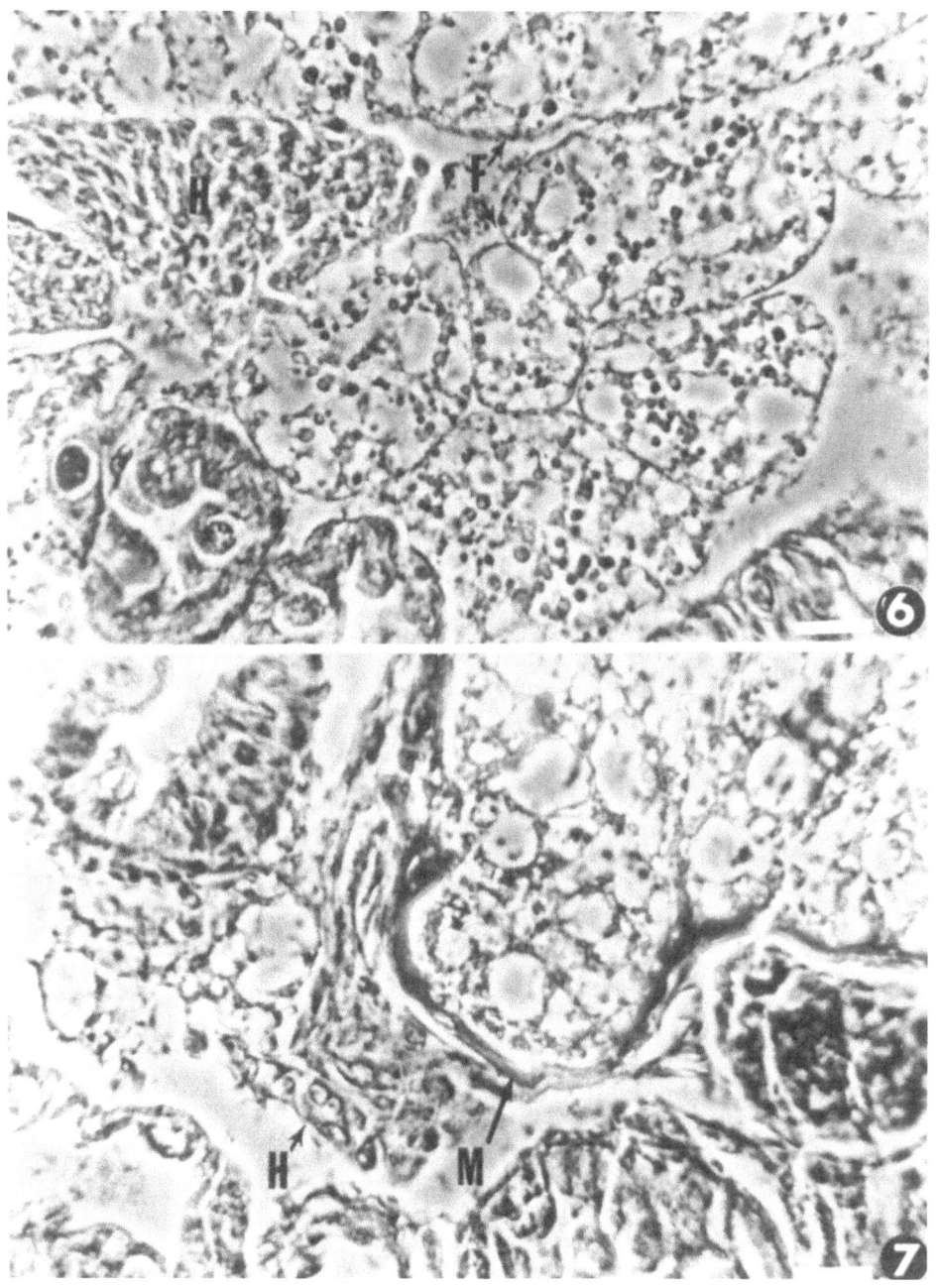

Fig. 8. Portion of a fat body cell from an early second-instar
         larva of vg tu^{28} showing evidence of deterioration.
         Note "empty" appearance of cytoplasm and areas where the
         plasma membrane has ruptured (arrows). One of two
         adjacent hemocytes shows signs of lysis. Mi, membrane-
         limited inclusions; Nu, hemocyte nuclei; L, lipid drop-
         lets; P, protein granule. X 8900. Scale, 1.0 μm.

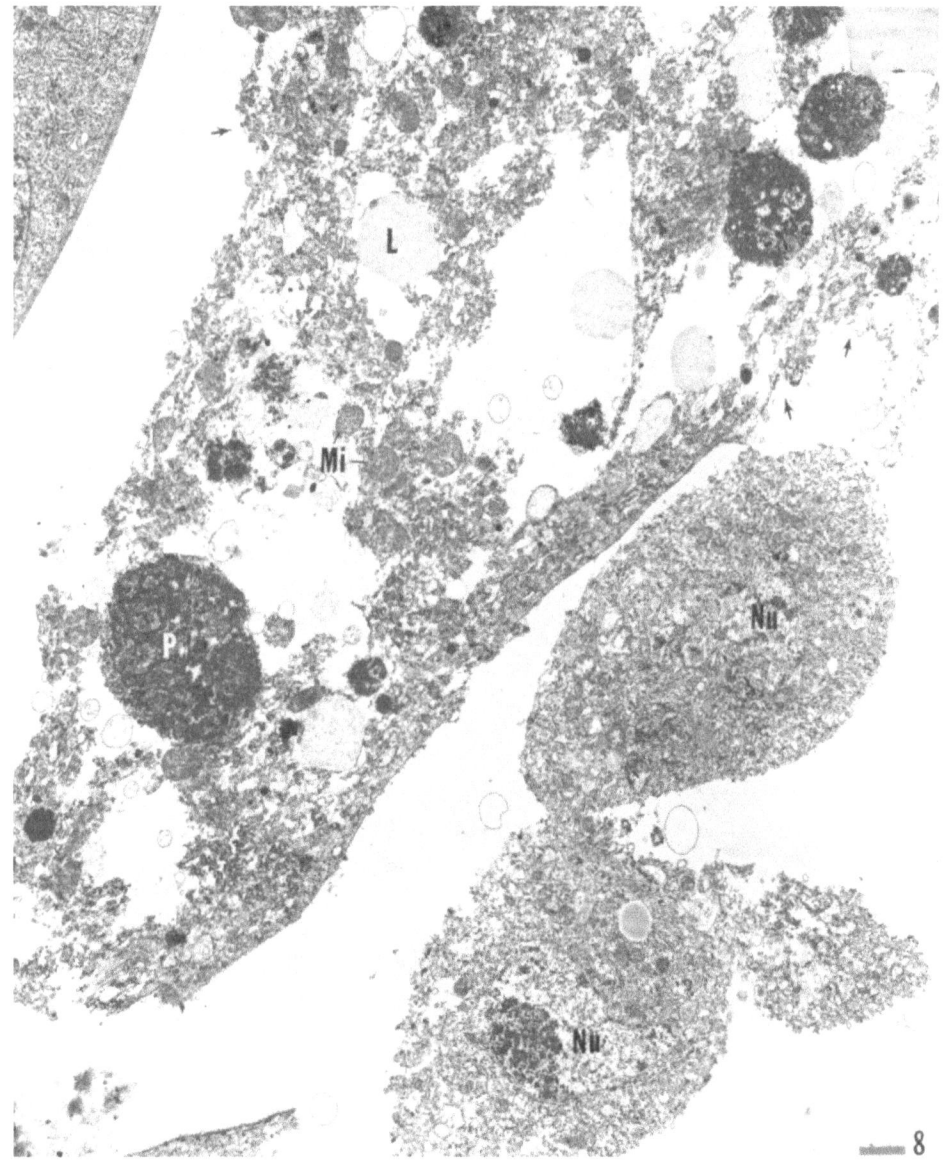

Fig. 9.   Portion of a protein granule (P), and cytoplasmic frag-
          ments (Cy) from a ruptured fat body cell near two intact
          hemocytes. Nu, nucleus of hemocyte. X 15,700. Scale,
          1.0 μm.

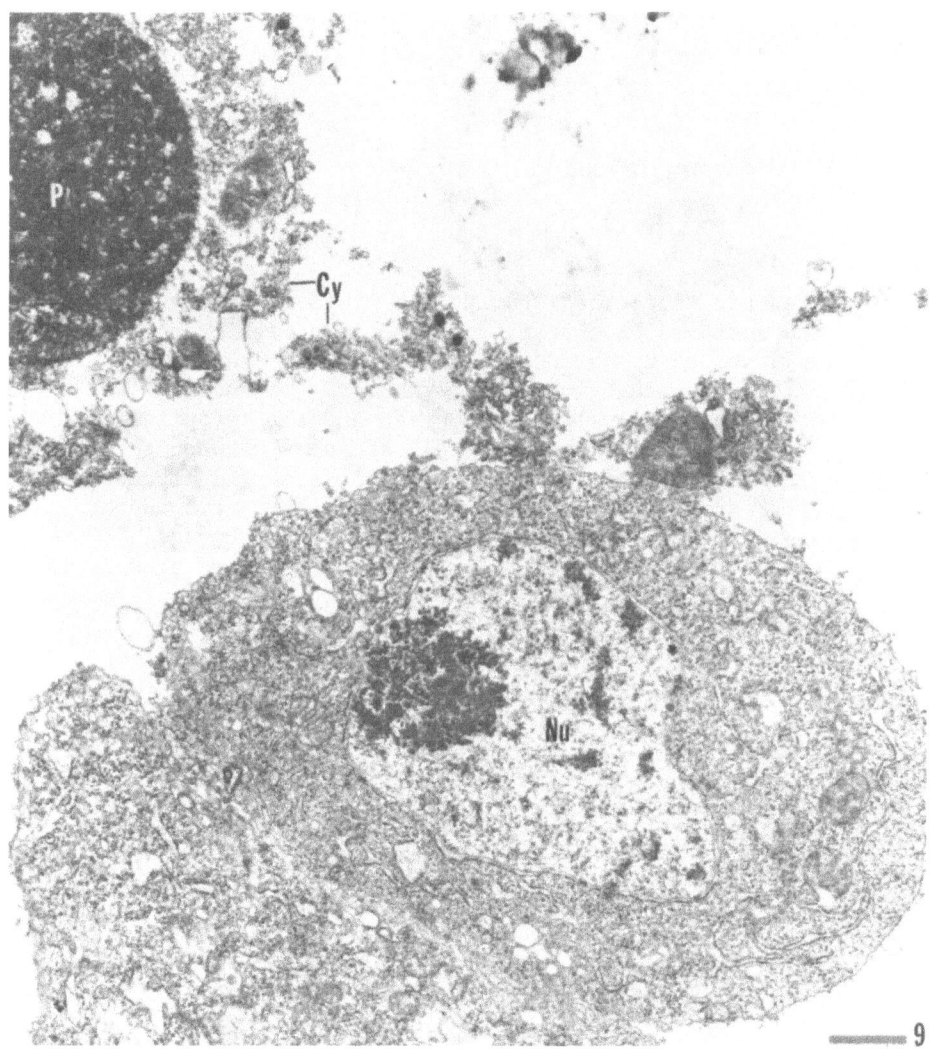

Fig. 10. Electron micrograph showing the lysis of hemocytes at
          the surfaces of abnormal, degenerating fat body cells
          of a second-instar $vg\ tu^{28}$ larva during the early
          stages of melanotic tumor formation. The nuclei (Nu)
          of two partially lysed hemocytes and the cytoplasmic
          fragments (Cy) of a third cell are seen between the
          plasma membranes (Pm) of two dissociated and degenerating
          fat body cells. Note the ruptured plasma membrane of
          one hemocyte (arrows). Lamellar bodies (Lb) and
          numerous membrane-limited inclusions (Mi) are present in
          both the fat cells and hemocytes. L, lipid droplets.
          X 11400 Scale, 1.0 μm.

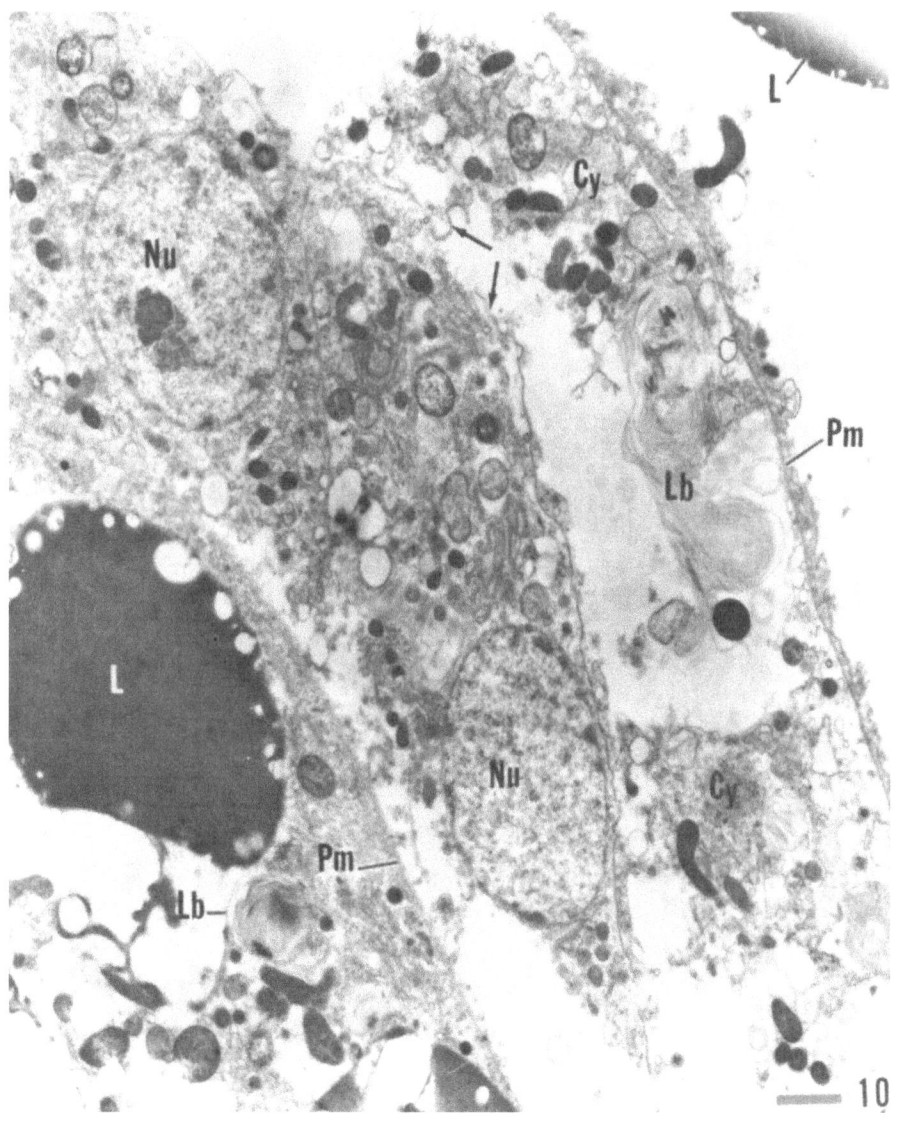

Fig. 11. Scanning electron micrograph of a developing melanotic
          tumor removed from the posterior hemocoel of a third-
          instar vg tu^{28} larva. Note the peripheral hemocytes
          adhering to the outer surface. X 450. Scale, 20 μm.

Fig. 12. Higher magnification of cells shown in Figure 11. Note
          small, spherical droplets on the surface of the
          encapsulated mass. X 4500. Scale, 2.0 μm.

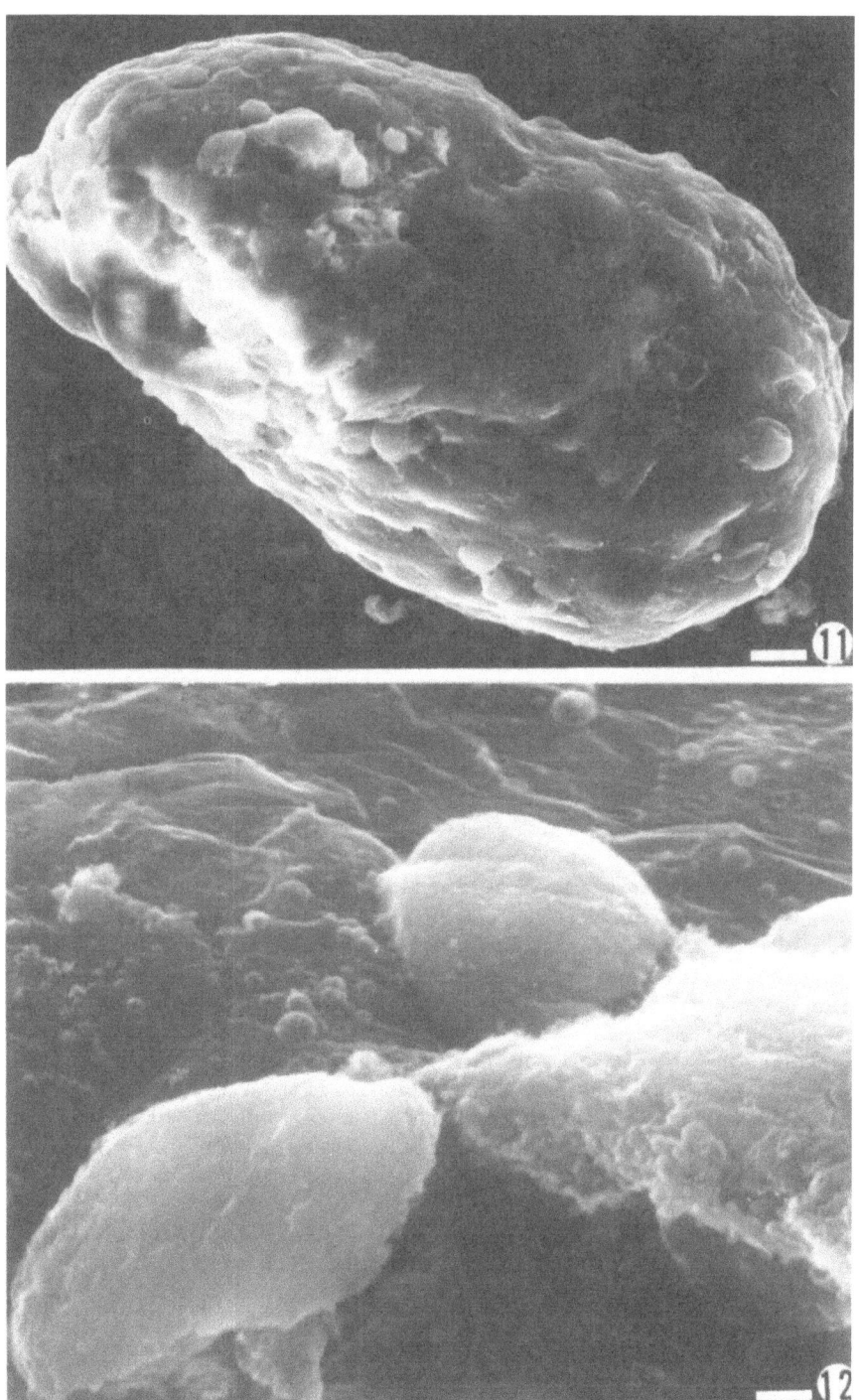

bodies, the accumulation of membrane-limited inclusions, and some cellular deterioration. The hemocytes involved in the reaction contained more lamellar bodies than did the fat body cells at the tumor forming sites. The early presence of cytoplasmic fragments in the extracellular spaces indicated that some fat body cells and at least the first hemocytes making contact with them had disrupted (Figs. 8-10). When completed, the tumorous mass was formed of several layers of hemocytes adhering closely to one another. Many of the cells, especially those of the innermost layers, were so heavily melanized that they could not easily be identified. The hemocytes making up the outer layer were less flattened than the cells below, and they closely resembled free hemocytes.

Scanning electron microscopy of the surfaces of encapsulated fat body cells in $vg\ tu^{28}$ larvae showed spherical particles or droplets similar to those observed by Rizki and Rizki (1974a,b) on tumorous masses in tu^W larvae (Figs. 11-13).

III. CELLULAR ENCAPSULATION OF PARASITES

Preliminary investigations at the ultrastructural level (unpubl.) showed that the hemocytes involved in the encapsulation of eggs of the wasp parasite *Pseudeucoila bochei* were primarily lamellocytes and some plasmatocytes. The cells aggregated and adhered to one another to form a capsule that completely enveloped the parasite about 2 days after infection. However, within 12 hr after parasitization there was extensive melanization of the chorion and underlying embryonic cells, evidence that the parasite was dead at this time. During the early stages of the host reaction the hemocytes at the surface of the parasite were characterized by many membrane-limited cytoplasmic inclusions, a large number of autophagic vacuoles, and numerous lamellar bodies (Figs. 14,15). None of the hemocytes examined possessed specialized organelles similar to vertebrate melanosomes. On and near the surface of the parasite was found a homogeneous layer of electron dense particles believed to be melanin. Smaller accumulations of these particles were found in the intercellular regions between some of the hemocytes forming the capsule (Figs. 16,17). Crystal cells, or oenocytoids, were not seen in the parasitized materials examined with the electron microscope. These larval hemocytes normally synthesize and/or concentrate phenols and enzymes associated with tyrosine metabolism (Rizki and Rizki, 1959).

Many of the cells near the parasite appeared necrotic with almost the entire cytoplasm abnormally electron dense and highly vacuolated. Some of the hemocytes formed extensive cell processes which insinuated among adjacent blood cells (Fig. 14), while other cells lysed. The definitive capsules were composed of an inner layer of melanin and melanized cellular components, a second region of lysed and necrotic hemocytes in various stages

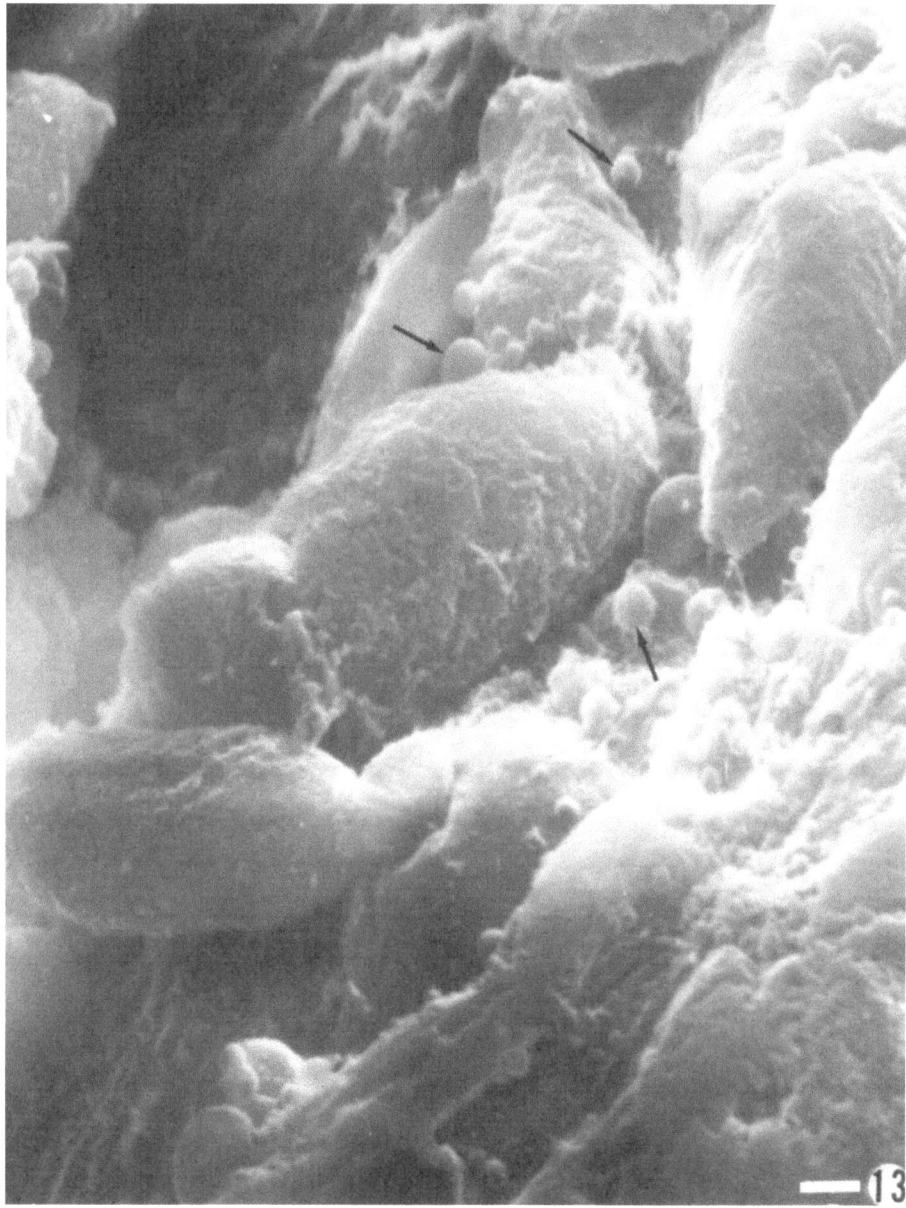

Fig. 13.   Scanning electron micrograph of a melanotic tumor
           taken from the posterior hemocoel of a second-instar
           vg tu[28] larva. Note spherical droplets (arrows) on and
           among the hemocytes. X 4500. Scale, 2.0 μm.

Fig. 14. Hemocytes involved in the encapsulation of *Pseudeucoila*
          bochei. Note lamellar bodies (Lb), autophagic vacuole
          (V), and numerous microvilli from two adjacent cells.
          Virus-like particles (arrows) are seen in the nucleus
          (Nu). Cy, cytoplasm of a second hemocyte which is
          shown in Figure 15. X 15000. Scale, 0.2 μm.

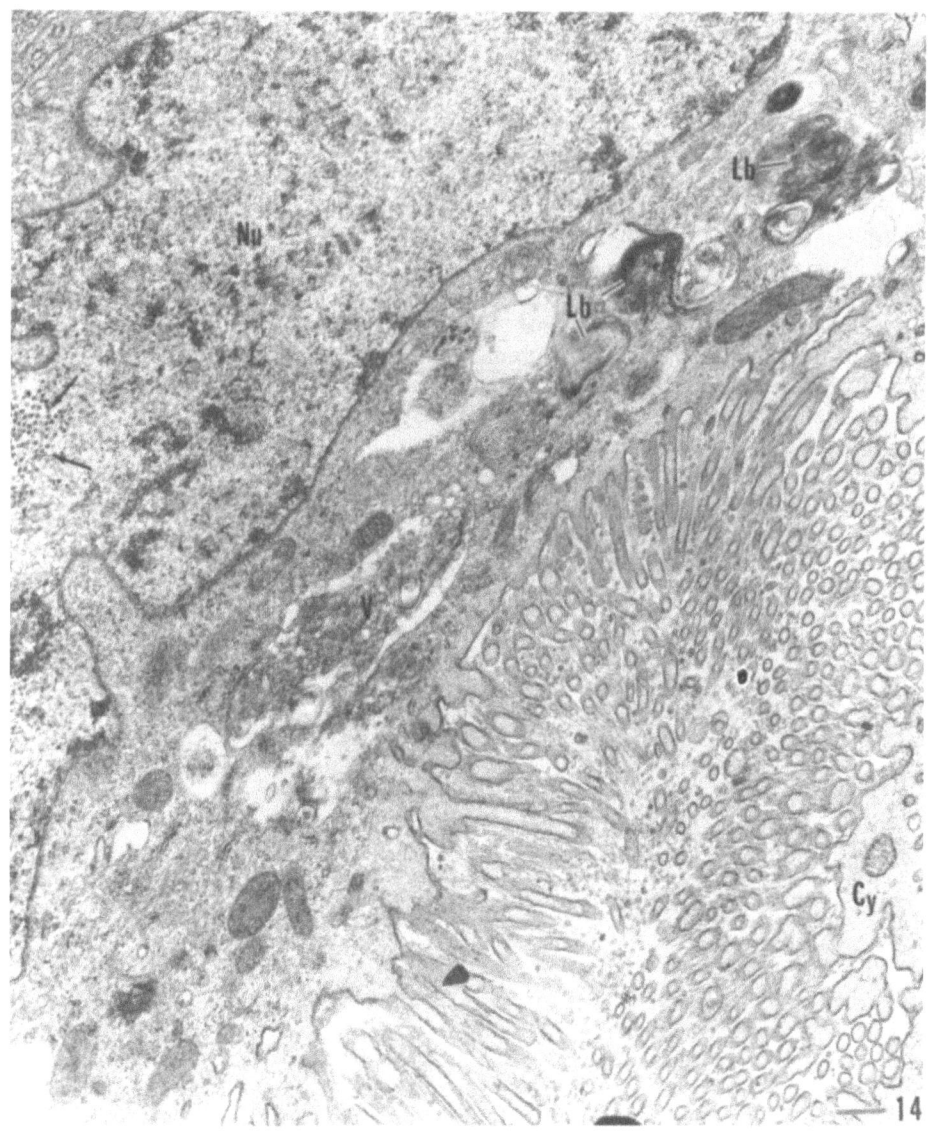

of pigmentation and degeneration, and a peripheral layer of closely
adhering hemocytes. The peripheral cells showed no loss of
structural integrity, and except for a few lamellar bodies and a
slightly more vacuolated cytoplasm, they resembled normal hemocytes
at a comparable stage of development (Fig. 18).

Examined with the scanning electron microscope, the surface of
the encapsulated parasite egg was comprised of irregularly layered
hemocytes. In addition, small, spherical particles were found
on and among the hemocytes which formed the capsule (Figs. 19-21).

IV. DISCUSSION AND CONCLUSIONS

In host larvae of *Drosophila* the abnormal fat body cells and
parasite eggs that were encapsulated by hemocytes showed pathological changes long before capsules were completely formed. In
both instances melanization was first noted in the peripheral
regions, along the plasma membranes of the fat body cells, and on
the chorion and underlying embryonic cells of the parasite. It
is interesting that the accumulation of numerous lamellar bodies
in host cells during parasite encapsulation and tumorigenesis
occurs also in aging *Drosophila* (Miquel *et al.*, 1972, 1974).
Electron microscopical investigations showed the most striking
feature in certain cells of old flies was the accumulation of
numerous lamellated cytoplasmic inclusions which resembled the
cytoplasmic bodies of human cercoid storage disease (Miquel
et al., 1972). Recent studies by Miquel *et al.* (1974) have demonstrated acid phosphatase activity in the lamellar bodies identifying these structures as lysosomes. Perhaps the pathological
changes occurring during host responses to foreign and abnormal
cells represent an acceleration of the normal aging process.
According to Casarett (1963), age-related diseases are essentially
either degenerative or neoplastic in character.

The pathological significance of *Drosophila* melanotic tumors
has long been questioned. Some investigators consider the abnormal
growths as merely inflammatory or degenerative manifestations,
whereas others feel that neoplastic processes are involved in
their formation (see Ghelelovitch, 1969; Harshbarger and Taylor,
1968). According to Harshbarger (1974), very few indisputable
neoplasms are known from insects and other invertebrates. Among
the salient characteristics of vertebrate neoplastic cells
apparently absent in cells comprising melanotic tumors are the
following: (1) Enhanced cell proliferation, (2) invasiveness,
(3) lethality to the host, (4) morphological and physiological
alterations of the cells, (5) loss of the ability to differentiate,
(6) instability of the karyotype, and (7) ability to be subcultured. However, most of the studies in the early literature
were descriptive accounts, lacking experimental data on whether
cells comprising melanotic tumors were rapidly proliferating and
capable of metastasizing to new foci and invading and destroying
host tissues.

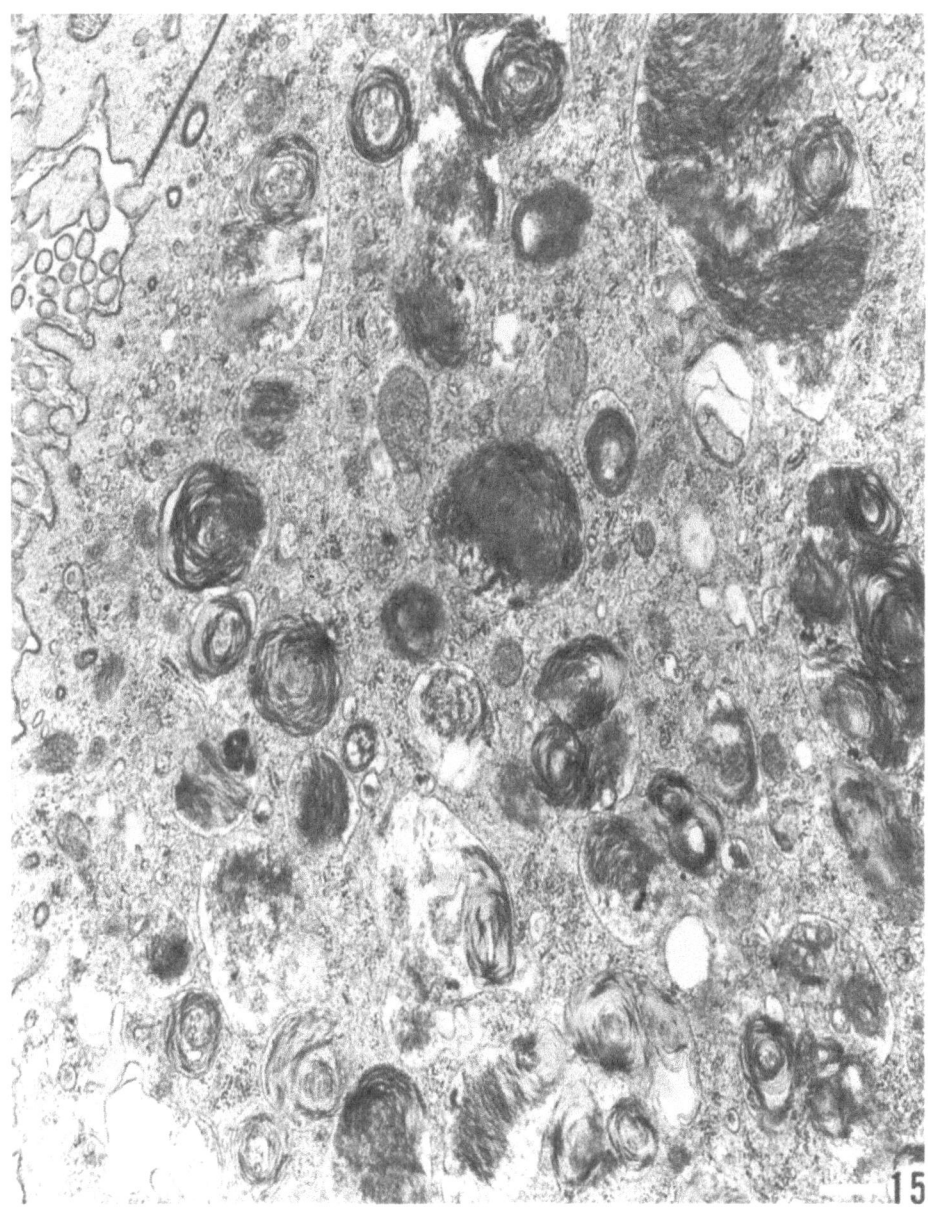

Fig. 15.   Hemocyte with numerous concentrically lamellated
           cytoplasmic inclusions (lamellar bodies). X 50000.
           Scale, 0.2 μm.

Fig. 16. Section through a hemocyte capsule developing around
          the egg of *Pseudeucoila bochei*. Note melanization
          (arrows) occurring along the plasma membranes of the
          adhering, necrotic cells, and the extracellular
          accumulations of electron dense particles (P) believed
          to be melanin. A portion of one cell at lower left
          has separated and folded back leaving a space just
          above a large accumulation of pigment. X 30000. Scale,
          0.5 μm.

Fig. 17. A high magnification of an accumulation of electron
          dense particles similar to those in Figure 16. A
          portion of a cell appears below and extends slightly
          into the particles. X 150000. Scale, 0.1 μm.

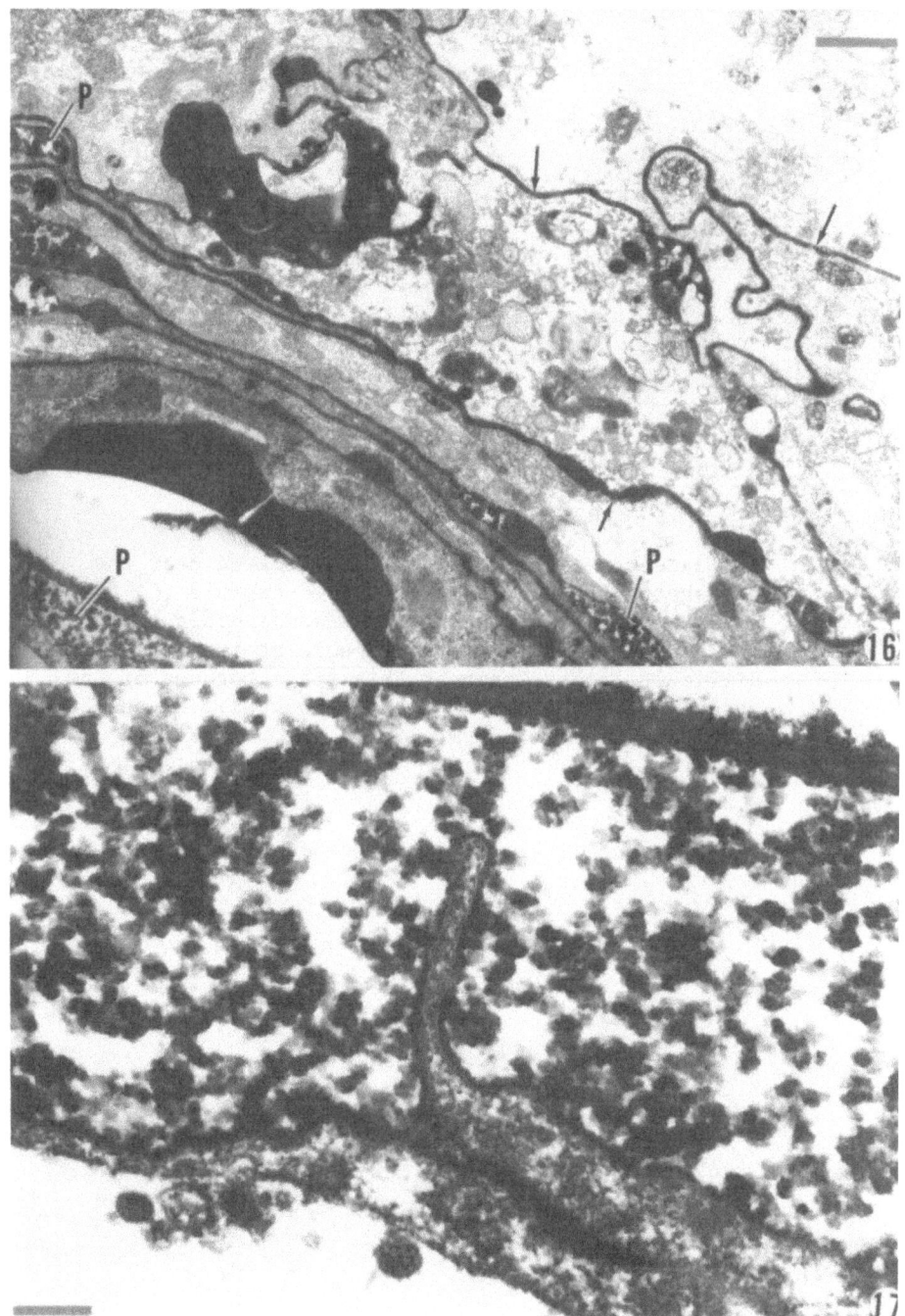

Fig. 18. A flattened, but intact hemocyte located near the
          periphery of a partially developed capsule (C) formed
          around the egg of the parasite *Pseudeucoila bochei*.
          The cellular nature of the capsule at the left is
          difficult to ascertain at this stage because of the
          extensive melanization of the hemocytes. Nu, nucleus.
          X 15500. Scale, 1.0 µm.

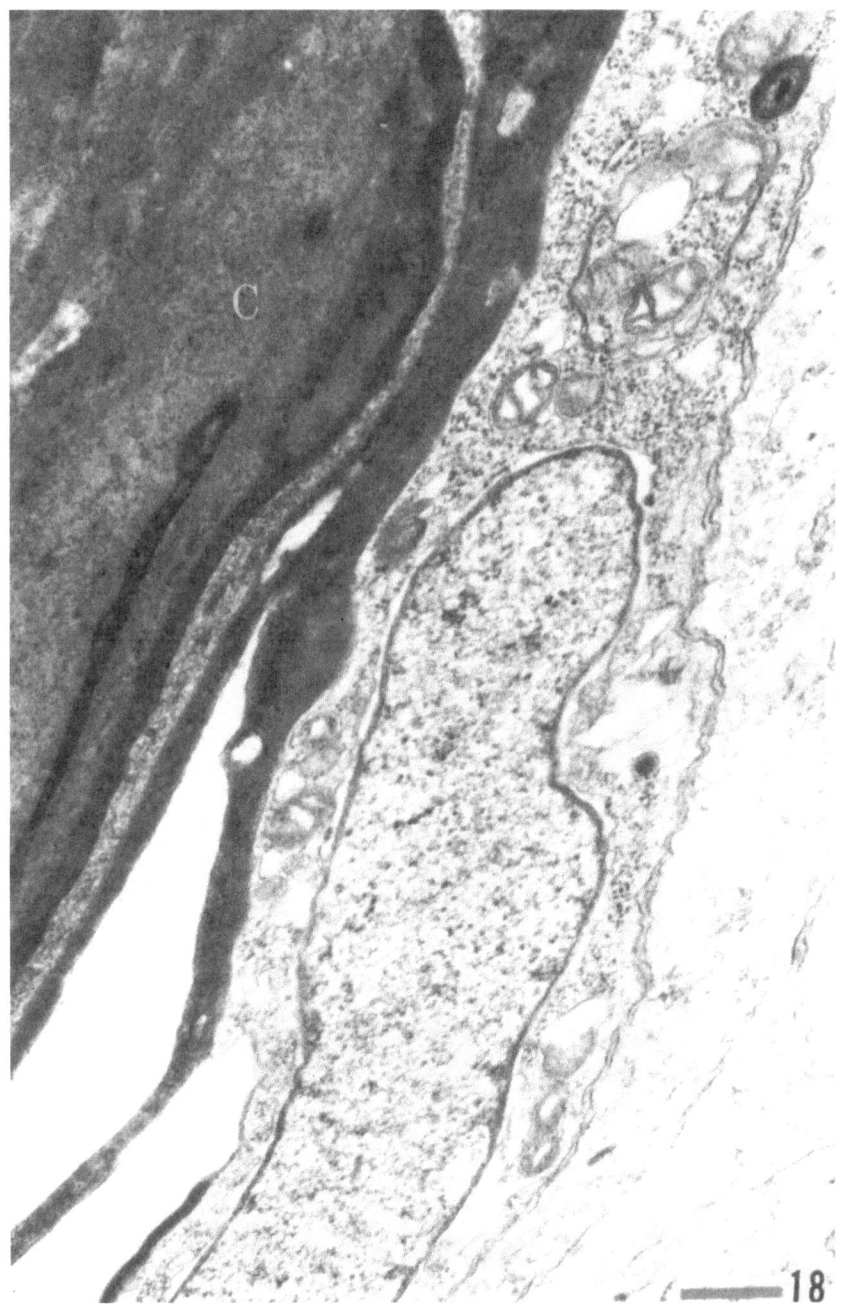

Fig. 19.   Scanning electron micrograph of an encapsulated egg (E)
           of *Pseudeucoila bochei* within the hemocoel of a third-
           instar host larvae of *Drosophila melanogaster*.   X 100.
           Scale 100 μm.

Fig. 20.   Higher magnification of the encapsulated egg in
           Figure 19, showing adhering hemocytes.   X 1000.   Scale,
           10 μm.

Fig. 21.   Scanning electron micrograph of the surface of a
           partially encapsulated egg removed from a second-instar
           larva showing small surface droplets or particles
           (arrows). X 1850.   Scale 10 μm.

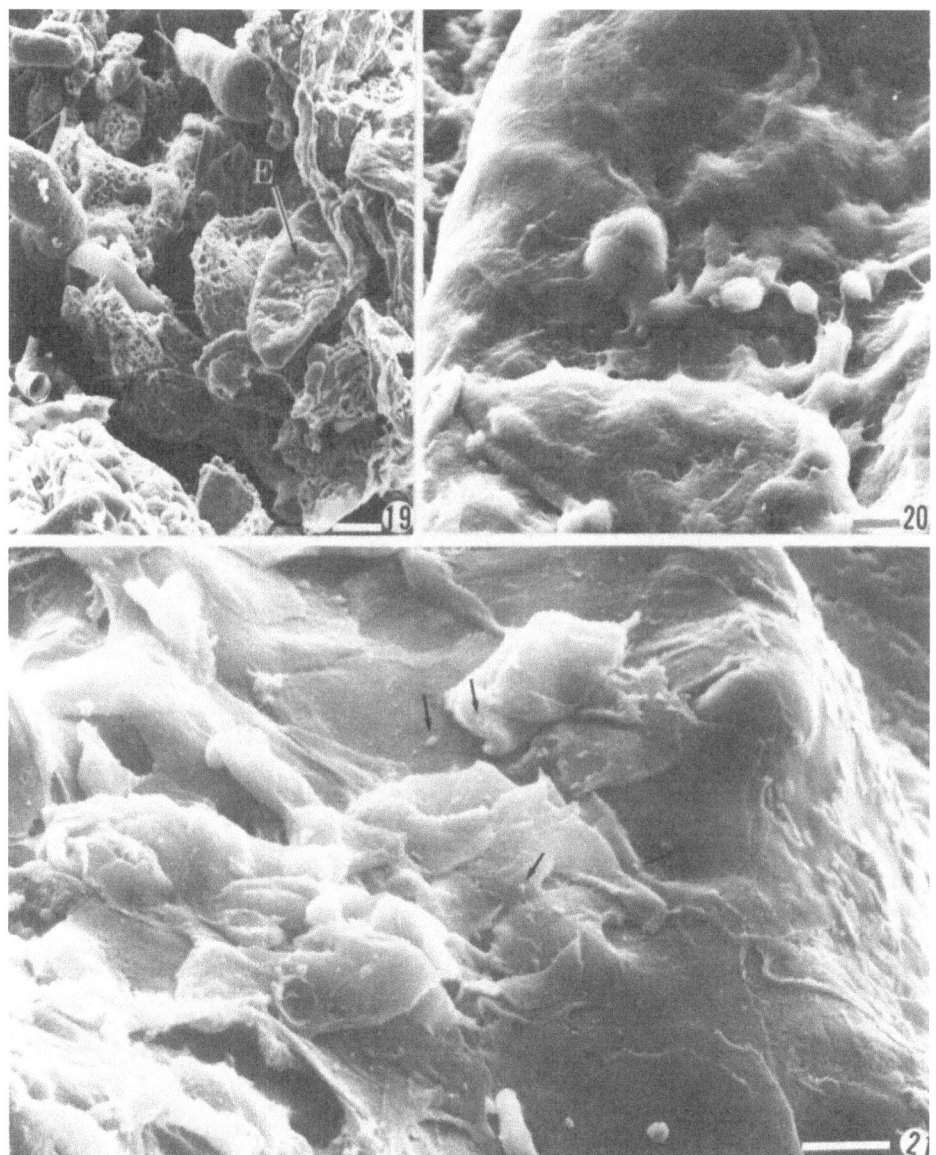

It has been suggested that the reason why malignancies are
absent or occur only rarely among invertebrates is because these
organisms do not carry the genetic capacity for neoplastic
transformations (Good and Finstad, 1969). It is maintained that
potentially neoplastic cells arise often during the development
of chordates, but the cell-mediated lymphoid systems of these
organisms are capable of discerning foreignness and eliminating
malignant cells deviant in surface antigenicity. Presumably
with such an efficient immune system against aberrant host cells,
vertebrates can tolerate a much greater genetic capacity for
neoplastic transformation than can insects or other invertebrates.

Despite marked differences in development between insects
and vertebrates, the ability of cells to become neoplastic is
believed to be dependent on the capacity of cells to divide and
on their state of differentiation (Gateff and Schneiderman, 1974).
Accordingly, as insects metamorphose there is a decrease in the
proportion of cells that are capable of dividing and of being
potentially neoplastic. In the larval stages only the hemocytes
and cells of the adult primordia divide, and only the hemocytes
and cells in the gonads and gut divide in the adult fly. Recent
studies by Gateff and Schneiderman (1959, 1974) and Hauri (1974)
have shown that true neoplastic transformations do occur in
Drosophila and are indeed closely linked with the capacity of
cells to divide.

Histologically, the cellular immune reaction of *Drosophila*
against *Pseudeucoila* was very similar to the melanotic encapsulation reaction reported by Poinar *et al.* (1968) for *Diabrotica*
beetles against juveniles of the mermithid *Filipjevimermis*
leipsandra. It is significant that eggs of *P. bochei* were dead
early during the immune response, at a time when melanization was
first noted, but long before the enveloping cellular capsules had
completely formed. These observations suggest that a type of
humoral response, perhaps associated with the melanization reaction,
occurred either before or during hemocyte contact with the surface
of the parasite. Unfortunately, it is not known whether melanin
is synthesized and released from host hemocytes directly, or if
the pigment forms from the noncellular components of the hemolymph.

In their study of tumor formation in tuB_3, Perotti and Bairati
(1968) were unable to determine the origin of melanic deposits,
but they suggested the pigment resulted from a spontaneous
transformation of the protein in the protoplasma due to intrinsic
or extrinsic agents. Perhaps disrupted hemocytes at the surface
of foreign objects and abnormal host cells release substances
which activate a polyphenol-phenoloxidase system in the hemolymph
and/or in other hemocytes. A type of humoral melanization, but
apparently not involving the participation of hemocytes, has been
reported by Götz (1969), Götz and Vey (1974), and Poinar and
Leutenegger (1971). This type of immune response was attributed

solely to the coagulation of hemolymph components directly on the surface of the parasite.

Very little is known about the stimulus causing insect hemocytes to aggregate and adhere to the foreign surface of a parasite and to abnormally developing host tissues. Recently, Vinson and Scott (1974) suggested that an electron dense extracellular substance prepared the surface of the egg of the braconid parasite, *Cardiochiles nigriceps*, for encapsulation in *Heliothis zea*. Unfortunately, the nature and origin of this extracellular substance which coats the eggs are not known. However, since the material was first noted on eggs of the parasite that had been already partially encapsulated, it is possible the substance was derived, at least in part, from ruptured hemocytes. In the tu^W mutant strain, Rizki and Rizki (1974a,b) considered the underlying factor in the formation of melanotic tumors to be the loss of the integrity of the basement membrane surrounding the fat body cells. In nontumorous larvae the basement membrane remained intact throughout larval development. Apparently disintegration of the membrane and the dissociation of fat cells normally occurred only after pupation. The authors suggested that the particulate material which forms around the dissociated fat body cells may be responsible for stimulating the hemocytes to aggregate, invade the affected areas, and encapsulate the abnormal cells. Similar particles were found on the encapsulated surfaces of both *P. bochei* eggs and abnormal fat body in $vg\ tu^{28}$ larvae. Thus, it would be of considerable interest to learn the origin of these particles, and to know if they appear during normal development.

ACKNOWLEDGMENTS

I am grateful to Drs. T. M. Rizki and R. Rizki for the scanning electron micrographs of tu^W melanotic tumors. I thank Dr. Terrence M. Hammill for helpful comments. I thank Dr. W. Côté, Mr. A. Day, and Mr. J. McKeon of the Electron Microscope Laboratory, SUNY at Syracuse for their cooperation and technical assistance. Studies by the author were supported by grants from the Research Foundation of the State University of New York, The National Institutes of Health, and the National Science Foundation

REFERENCES

Bairati, A., Jr., and M. E. Perotti (1966). Electron microscope studies of the ultrastructure of melanotic tumors in *Drosophila melanogaster*. *In* "Atti V. Congr. It. Microscopy Elettron." pp. 42-45. Tipogr. Seminario, Padova.

Casarett, G. W. (1963). Concept and criteria of radiologic aging. *In* "Cellular Basis and Aetiology of Late Somatic Effects of Ionizing Radiation." (R.J.C. Harris, ed.), pp. 189-205. Academic Press, New York.

Castiglioni, M. C. (1957). Le cellule dell'emolinfa di *Drosophila melanogaster* in relazione al genotipo e alla produzione di pseudotumori. *Atti III Riun. Assoc. Genet. Ital. La Ric. Sci.*, suppl., 27, 51-57.

Gateff, E., and H. A. Schneiderman (1969). Neoplasms in mutant and cultured wild-type tissues of *Drosophila. Nat. Cancer Inst. Monogr.*, 31, 365-397.

Gateff, E., and H. A. Schneiderman (1974). Developmental capacities of benign and malignant neoplasms of *Drosophila. Wilhelm Roux' Archiv.*, 176, 23-65.

Ghelelovitch, S. (1969). Melanotic tumors in *Drosophila melanogaster. Nat. Cancer Inst. Monogr.*, 31, 263-275.

Good, R. A., and J. Finstad (1969). Essential relationship between the lymphoid system, immunity, and malignancy. *Nat. Cancer Inst. Monogr.*, 31, 41-58.

Götz, P. (1969). Die Einkapselung von Parasiten in der Hamolymphe von Chironomus-Larven (Diptera). *Zool. Anz., Suppl.*, 33, 610-617.

Götz, P., and A. Vey (1974). Humoral encapsulation in Diptera (Insecta): defence reactions of *Chironomus* larvae against fungi. *Parasitology*, 68.

Harshbarger, J. C. (1974). Radiation, neoplasms, carcinogenic chemicals, and insects. *In* "Insect Diseases." (G. E. Cantwell, ed.), Vol. 2, pp. 377-416. Marcel Dekker, Inc. New York.

Harshbarger, J. C., and R. L. Taylor (1968). Neoplasms of insects. *Ann. Rev. Ent.*, 13, 159-190.

Hauri, H. (1974). Ein invasives Neoplasms aus embryonalen Zellen von *Drosophila melanogaster* in Dauerkultur *in vivo. J. Embryol. Exp. Morph.*, 31, 347-375.

Miquel, J., K. G. Bensch, D. E. Philpott, and H. Atlan (1972). Natural aging and radiation-induced life shortning in *Drosophila melanogaster. Mech. Age. Dev.*, 1, 71-97.

Miquel, J., A. L. Tappel, C. J. Dillard, M. M. Herman, and K. G. Bensch (1974). Fluorescent products and lysosomal components in aging *Drosophila melanogaster. J. Gerontol.*, 6, 622-637.

Nappi, A. J. (1974). Insect hemocytes and the problem of host recognition of foreignness. *In* "Contemporary Topics in Immunobiology." (E. L. Cooper, ed.), Vol. 4, pp. 207-224. Plenum Press, New York.

Nappi, A. J. (1975a). Parasite encapsulation in insects. *In* "Invertebrate Immunity. Mechanisms of Invertebrate Vector-Parasite Relations." (K. Maramorosch and R. E. Shope, eds.), pp. 293-326. Academic Press, New York.

Nappi, A. J. (1975b). Cellular immune reactions of larvae of *Drosophila algonquin*. *Parasitology*, 70, 189-194.

Nappi, A. J. (1975c). Inhibition by parasites of melanotic tumor formation in *Drosophila melanogaster*. *Nature*, 255, 404-404.

Nappi, A. J., and F. A. Streams (1969). Haemocytic reactions of *Drosophila melanogaster* to the parasites *Pseudeucoila mellipes* and *P. bochei*. *J. Insect Physiol.*, 15, 1551-1566.

Oftedal, P. (1953). The histogenesis of a new tumor in *Drosophila melanogaster*, and a comparison with tumors of five other stocks. *Z. Indukt. Abstammungs. Vererbungsl.*, 85, 408-422.

Perotti, M. E. and A. Bairati, Jr. (1968). Ultrastructure of the melanotic masses in two tumorous strains of *Drosophila melanogaster (tuB_3* and *Freckled)*. *J. Invertebr. Pathol.*, 10, 122-138.

Poinar, G. O., Jr. (1969). Arthropod immunity to worms. *In* "Immunity to Parasitic Animals." (G. J. Jackson, R. Herman, and I. Singer, eds.), Vol. 1, pp. 173-210. Appleton-Century-Crofts, New York.

Poinar, G. O., Jr. (1974). Insect immunity to parasitic nematodes. *In* "Contemporary Topics in Immunobiology." (E. L. Cooper, ed.), Vol. 4, pp. 167-178. Plenum Press, New York.

Poinar, G. O., Jr., and R. Leutennegger (1971). Ultrastructural investigations of the melanization process in *Culex pipiens* (Culicidae) in response to a parasitic nematode (Mermithidae). *J. Ultrastruct. Res.*, 36, 149-158.

Poinar, G. O., Jr., R. Leutenegger, and P. Götz (1968). Ultrastructure of the formation of a melanotic capsule in *Diabrotica* (Coleoptera) in response to a parasitic namatode (Mermithidae). *J. Ultrastruct. Res.*, 25, 293-306.

Rizki, T. M. (1957). Tumor formation in relation to metamorphosis in *Drosophila melanogaster*. *J. Morph.*, 100, 459-472.

Rizki, T. M. (1960). Melanotic tumor formation in *Drosophila*. *J. Morph.*, 106, 147-157.

Rizki, T. M., and R. M. Rizki (1959). Functional significance of the crystal cells in the larva of *Drosophila melanogaster*. *J. Biophys. Cytol.*, 5, 235-240.

Rizki, T. M., and R. M. Rizki (1974a). Topology of the caudal fat body of the tumor[W] mutant of *Drosophila melanogaster*. *J. Invertebr. Pathol.*, 24, 37-40.

Rizki, T. M., and R. M. Rizki (1974b). Basement membrane abnormalities in melanotic tumor formation of *Drosophila*. *Experientia*, 30, 543-546.

Salt, G. (1963). The defense reactions of insects to metazoan parasites. *Parasitology*, 53, 527-642.

Salt, G. (1970). "The Cellular Defense Reactions of Insects." Cambridge University Press, London.

Scharrer, B. and M. S. Lochhead. Tumors in the invertebrates: A review. *Cancer Res.*, 10, 403-419.

Stark, M. B. (1918). An hereditary tumor in the fruit fly,
    Drosophila. *J. Cancer Res.*, <u>3</u>, 279-301.
Stark, M. B. (1919). A benign tumor that is hereditary in
    Drosophila. *Proc. Nat. Acad. Sci.*, <u>5</u>, 578-580.
Vinson, S. B., and J. R. Scott (1974). Parasitoid egg shell
    changes in a suitable and unsuitable host. *J. Ultrastruct.
    Res.*, <u>47</u>, 1-15.
Walker, I. (1959). Die Abwehrreaktion des Wirtes *Drosophila
    melanogaster* gegen die zoophage Cynipidae *Pseudeucoila bochei*
    Weld. *Rev. Suisse Zool.*, <u>68</u>, 569-632.
Whitcomb, R. F., M. Shapiro, and R. R. Granados (1974).
    Insect defense mechanisms against microorganisms and parasi-
    toids. *In* "The Physiology of Insecta." (M. Rockstein, ed.)
    Vol. 5. Academic Press, New York.

INDEX